JAPANESE PLANTS
Know Them & Use Them

Betty W. Richards
B. Sc. (Hort.)
Anne Kaneko
B. A. (Hons. Japanese)

SHUFUNOTOMO CO., LTD.
Tokyo Japan

Second printing, 1989

All rights reserved throughout the world.
No part of this book may be reproduced in any form
without permission in writing from the publisher.

© Copyright in Japan 1988 by Betty W. Richards &
                             Anne Kaneko
Photographs by Betty W. Richards & Anne Kaneko
              Ars Photo Planning
              Naoki Uehara
Book Design by Momoyo Nishimura

Published by Shufunotomo Co., Ltd.
2-9, Kanda Surugadai, Chiyoda-ku, Tokyo, 101 Japan

ISBN: 4-07-975121-4
Printed in Japan

CONTENTS

Introduction —————————— 5

Japanese Plants ————————— 7

Appendix I

    Botanic Gardens ——————— 209

    Other Gardens in Tokyo ———— 213

    Gardens in Kyoto —————— 215

    Plant Fairs and Exhibitions in Tokyo - 215

Appendix II

    The Seven Herbs of Spring ——— 217

    The Seven Flowers of Autumn —— 217

Appendix III

    Classification of Cherry Blossoms — 219

Bibliography ————————— 221

Index ————————————— 222

INTRODUCTION

There are few flowering plants in a typical Japanese landscape garden, but even so, the Japanese have been hybridising many plants for centuries; chrysanthemums, cherries, irises and azaleas coming to mind immediately. They use them to brighten up balconies and odd corners as well as putting them in strategic positions in the gardens. Their attention to detail, appreciation of minutiae, along with the harmony, grace and refinement felt in all their art forms, has given a great heritage. We attempt to show some of this to you, not only so that you may become more knowledgeable botanically, but also so that you may better appreciate a long tradition and culture. You will be reminded, too, of the great plant hunters of the past who gave their names to so many of the native plants and introduced them to Europe—Forrest, Veitch, von Siebold, Fortune, to name but a few. Then there are all the "japonica's", and "nipponense's", luxuriant in their native habitats. Where would our own gardens be without them?

As a first time visitor, you will probably not get much further than the Kanto (Tokyo area) and Kansai (Kyoto-Osaka area) with their mainly warm-temperate climates. Throughout Japan, the climate extends from sub-tropical—with more southerly latitudes and greater protection (e.g. in heavily built up areas), to cool-temperate—with more northerly latitudes or with elevation in the mountain areas. Ideally you should start your visit when

the cherries are in flower and extend it until the azaleas are out—or plan a trip in mid-May from Hokkaido to Kyoto when you may catch cherry blossoms and azaleas in one week. Then come back in the clear blue days of autumn for the blazing colours of the leaves, and for the chrysanthemums, leaving the irises and lilies of sticky summer until you are so entranced by the country and the people that you can put up with the climate.

This cannot be an exhaustive work but we have made every effort to be botanically correct. In this respect we are particularly indebted to Professor Ryuzo Sakiyama of the Department of Horticulture, Faculty of Agriculture, Tokyo University who not only generously agreed to check the manuscript but also enrolled the help of colleagues in various fields. We would like to express here our sincere appreciation to these people: Professor Hiroshi Tanaka, of the Faculty of Agriculture, Tamagawa University; Dr. Masao Yoshida, Director of the Ministry of Agriculture, Forestry and Fisheries' Fruit Tree Research Station in Morioka; Professor Shuichi Iwahori, of the Faculty of Agriculture, Kagoshima University; and Dr. Ryosuke Ogata, Associate Professor of the Faculty of Agriculture, Tohoku University.

JAPANESE PLANTS

ADONIS

Adonis amurensis   (Ranunculaceae)   ***(Fukujusō)***

A dwarf native perennial of 7–10 cm seen especially in cooler areas. Grown in gardens and in pots for its early flowers. Finely cut dark green leaves, silver grey buds, wide-open yellow flowers 3–4 cm diameter with up to 30 glossy petals. Some apricot and pink forms are known.

The Japanese name means "wealth and happiness", and the bright flowers bursting through the bare earth and snow really do bring joy at the coldest time of the year. They are often used in mixed, potted arrangements for the New Year.

An Ainu legend tells the story of a Goddess turned into this plant when she refused to marry the god of her father's choice. This is surprisingly similar to the Greek legend of the death of Adonis, from whose blood blossomed a plant—the red ***Adonis annua*** of the Mediterranean area.

FUKUJUSŌ

APPLES & CRAB APPLES

Eating Apples *(Ringo)*
Malus pumila domestica   (Rosaceae)

An important crop of the cooler parts, especially Aomori. Trees are flat crowned often with the head at 3–4 m and drooping, spreading branches. In modern apple orchards a dwarfing rootstock is used and the trees are grown on the central leader system. Pink and white blossoms in May are thinned by hand to encourage very large dessert apples up tp 12 cm or more across. **Fuji**, the leading variety, originated in Japan and occupies nearly 50% of total production. All varieties are sweet, juicy and flavoursome, intended to be served peeled and sliced on small delicate plates. Sharp cooking apples are not available but the small, red **Kōgyoku** makes a good apple pie.

Flowering Crab Apple *(Hana-kaidō)*
Malus halliana

A native of China, long grown in gardens, making a small (1.5–4 m) crown with spreading branches, full of pink flowers in mid-April. The flowers are very delicate, semi-double blossoms which hang on long stalks. The outsides of the petals are darker than the insides. In autumn the tiny apples (5–9 mm) may be yellow or a dark reddish brown, and these too hang down most attractively. All these characteristics make it a good bonsai subject and a popular garden plant.

RINGO

HANA-KAIDŌ

APRICOTS AND PLUMS

Mume *(Ume)*

Prunus mume (Rosaceae)

Westerners who know the Apricot as a downy skinned, yellow or orange fruit, and a plum as a smooth skinned, yellow, green or red fruit, have had a long standing dilemma about the English word to use for *Prunus mume*. The most helpful way seems to be to stick to the word **Mume** for the green to yellow downy fruit and the flowering trees, to use Apricot for *Prunus armenaica*, and Japanese Plum for *Prunus salicina*. **Ume** would then be kept for the Japanese usage—as in **ume-boshi** and **ume-shu**. This is what we try to do below.

The type of **mume** grown for flowers has developed from the wild **mume** which in Feb/March, has single white or pale pink flowers of an intense, sweet fragrance, clustered close to shiny dark brown branches. Hundreds of cultivars have evolved through the centuries. The Japanese have classified these into 3 main groups according to the colour of the wood and into 8 sub-groups according to the size and shape of the leaves and the colour and form of the flowers (white through pink to red, and single or double). **Mume** flower outside in the very early spring but as bonsai they are encouraged to flower for the New Year, when the delicate pink blooms on the gnarled old trees are a promise of hopes and joys to come. Bonsai are usually from the Wild **Mume** and the Naniwa groups. The Bungo group are crossed with the Apricot, and are larger.

Whether in gardens or pots, they are all long lived trees, show their age with grace, and are much venerated. **Mume** are the subject of the first flower viewing of the year, when the frost is warmed by the promise of spring. But it is still too cold for picnicking under the trees. The 'viewing' is not so exuberant as at Cherry Blossom time, as befits the flower which is the patron-flower of scholars and men of letters. At Zuiganji Temple in Matsushima, two 400 year old **mume** trees, one red and one

MUME

MUME

white, flank the entrance. (Red and white is always considered a happy and joyful combination in Japan.)

Fruiting **mume** have been the subject of much research and development as they are an important food crop. The fruits when ripe are up to 3 cm, oval to round, downy, green to yellow, with a ridge at the side and a hard stone in the middle. The fruit, which is very sour, is dried in the sun in June, then preserved in salt and red **shiso** (*perilla frutescens*) to make the red pickled **mume** or **ume-boshi** which are put in the middle of rice balls or packed lunches (**bentō**) to keep the rice fresh. Pickled **mume** are also taken with green tea first thing in the morning. They are not as pleasant to Western tastes as is **ume-shu**, a wine similar in taste to cherry brandy, and not to be missed. This is made by putting green mume into distilled 40% alcohol (**shōchū**) with sugar.

Apricot *(Anzu)*

Prunus armeniaca

A native of China, long grown in Japan. The fruit is sour, yellow to red, with a downy skin and a rough, loose stone. It enjoys the cooler parts of the country and is being used in a breeding programme along with *P. mume* and European and N. American varieties of apricot, and *Prunus salicina*, to develop a dessert fruit that is disease resistant and climatically suited to Japan.

The Japanese Plum *(Sumomo)*

Prunus salicina

In the wild this makes a small bushy tree or a large shrub with white flowers crowding the shiny leafless branches early in the year but after *P. mume* has flowered. Its leaves turn vivid red in the autumn. A forefather of many varieties of the domestic plum.

MUME

Ume-shu and two kinds of ume-boshi.

ARDISIA

Coral Berry *(Manryō)*
Ardisia crenata (Myrsinaceae)

A low evergreen of 30–60 cm which grows wild in the woods in warmer areas but is usually seen as a pot plant. The leaves are leathery, dark green, pointed ovals of 7–12 cm, recognisable by the rounded toothing on their edges. The **Manryō** flowers in summer; short stalked bunches of starry white blossoms. In autumn bunches of stalked red berries appear, which stay on the bushes until the spring.

As a pot plant it is usually grown to have erect stems with the leaves concentrated at the tops of the canes and the prominent fruits hanging down in most attractive clusters. A '**ryō**' was a golden coin, and **Manryō** means 'ten thousand **ryō**', so it is a plant that has very lucky associations, especially for New Year. There is also a pale-yellow berried cultivar.

See also **Senryō** (*Chloranthus glaber*), page 176.

AUCUBA

Japanese Aucuba, Spotted Laurels *(Aoki)*
Aucuba japonica (Cornaceae)

Bushy evergreen shrubs to 1–3 m, growing wild in shady mountain areas. The leaves are leathery, 10–15 cm long, entire and slightly toothed. There are many garden varieties, classified and named according to the markings on the leaves.

Flowers are inconspicuous and appear on both male and female plants but the females have decorative clusters of red berries (1–1.5 cm) which persist through the winter. A very useful evergreen, tolerant of shade, pollution and clipping.

MANRYŌ

AOKI

AZALEAS

Rhododendron species and hybrids (Ericaceae)
(Tsutsuji, Satsuki, Shakunage)

All Azaleas are now classified as ***Rhododendron spp***. The species named below are chosen out of many Japanese native species because of their interest to Western growers. The two main hybrid groups in Japan are known as **Tsutsuji** and **Satsuki**.

Tsutsuji. Flowers at the end of April and into May, with 3–7 cm flowers, often hose-in-hose and semi-double. Can be seen everywhere—parks, roadsides, gardens.

Satsuki. Flowers in early June on low growing spreading bushes. The flowers on exhibition plants may be up to 8 cm diameter and are usually single. The parentage is confused but probably includes *R. indicum* and *R. eriocarpum*. A very 'sporty' group, anything may happen from a packet of seeds, and even an established plant can become a veritable 'family tree', with different branches bearing different coloured flowers. '**Satsuki**' refers to June, the 5th month of the old Japanese calendar, when this shrub flowers. Shows and exhibitions are held throughout the country.

R. Kaempferi **(Yama-tsutsuji)**

Semi-deciduous, 1–3 cm high, the wild azalea of the river banks and hillsides. Pale vermilion flowers, 4–5 cm, against fresh green leaves which turn crimson purple in the autumn. There are also variants with red and purple flowers, and infrequently, white ones.

SATSUKI

YAMA-TSUTSUJI

R. japonicum   (=Azalea mollis)　　　**(Renge-tsutsuji)**

Deciduous loose growing shrub, 1–2 m. Big clusters of scented, orange or yellow flowers each 5–7 cm. Rich autumn tints in the leaves. A parent of the Mollis Hybrids and Occidentale groups of deciduous azaleas and occuring in the genealogy of many other groups.

The following are evergreen azaleas found in Japan. They are mostly less than 1 m high, the height depending on the location in the wild or on the cultivator in the garden. They came from the warmer southern parts and have flowers from white, through pink to reddish purple.

R. indicum　　　**(Satsuki)**

A native of Japan in spite of its specific name.

R. obtusum　　　**(Kirishima)**

R. kiusianum　　　**(Miyama kirishima)**

Kurume tsutsuji, the most popular **tsutsuji**, is a hybrid of R. obtusum and R. sataense (the Sata-**tsutsuji**). Kurume is a town in the north of Kyushu where much of the initial hybridizing took place.

Rhododendron must be one of the most profligate genera in nature, hybridising readily both in the wild and in cultivation. We know that these azaleas were being brought into Europe in the 17th and 18th centuries by those intrepid seafarers the "East India Men", when "The Indies" was a vague unknown area to European botanists and Japan was almost closed to traders. Not surprising, then, that the Japanese R. indicum was so called and that these plants became known in England as the Belgian Indicas. Spectacular, compact, easily carried and

RENGE TSUTSUJI

MIYAMA KIRISHIMA

transplanted to grace the conservatories of the European aristocracy.

Meanwhile, behind its closed ports, the Japanese hybridisers were hard at work aiming at new and perfect varieties, particularly for bonsai and ikebana. They were using the **Mochitsu-tsuji** with its sticky calyx and the wild **Yama-tsutsuji**. By 1681 there was already a record of 150 garden varieties, mostly dwarfed plants with small leaves and tightly packed flowers. Florence Cane (1908) tells us about the **Satsuki** of Matsushima, the pine clad islands near Sendai, in Tohoku, 'wild, crimson, dwarfed by the climate of the islands and forming a ground cover only 6" high'*. E.H. Wilson was excited to find the ancestors of the Kurume Azaleas on thin volcanic soils 100 m above sea level on Mt. Kirishima in Kyushu in 1918. He considered the Kurumes to be his most exciting plant introduction into Europe and America and indeed they revolutionised spring gardens in cooler parts of Europe and North America with their spectacular colour. The Kurumes were the result of centuries of hybridizing by Japanese experts in Kyushu and, in the West, they were the start of yet another explosion of plant breeding.

The hybrid evergreen azaleas of Japanese gardens and parks are clipped and disciplined. Out of season, to Western eyes, it seems unlikely that these highly formalised bushes will flower. But flower they do. They erupt into cushions of all the colours of the sunset.

See them in Shinjuku-gyoen in Tokyo where azalea plantations first delighted the nobility of the Tokugawa period and now give pleasure to the people of Tokyo and their visitors. See them wild on the hillsides in the Nikkō-Chūzenji area.

The Flowers and Gardens of Japan, p. 88 by Florence du Cane

TSUTSUJI

SATSUKI "OSAKAZUKI"

BAMBOOS

Species of *Phyllostachys, Arundinaria, Sasa,* etc.
(Bambusaceae)

(Take, Sasa, etc.)

Bamboos have been the subject of much botanical research since World War II, and the nomenclature is constantly changing. The following is intended as a guide. The serious student of bamboos is recommended to see the extensive collection in the Kyoto Botanic Garden and to find a friend who reads Japanese to interpret the floras and literature about them.

Phyllostachys bambusoides (=*P reticulata*) *(Ma-dake)*

Height 15–20 m with a diameter of 15 cm and 2 prominent rings at the nodes. The most widely spread species in Japan.

Phyllostachys pubescens (=*p mitis*) *(Mōsō-chiku)*

Height 15–20 m with a diameter of up to 20 cm. The largest bamboo in Japan, especially on the red soils of the warmer South West. Node not prominent, with single ring only. The best shoots for food.

Black Bamboo
Phyllostachys nigra *(Kuro-chiku)*

2–8 m. Mature culms black. Very elegant and often used in gardens.

Sasa veitchii (=*Arundinaria veitchii*) *(Kuma-zasa)*

1–1.5 m. Small, hardy, one long branch per node. White edges to the leaves in winter.

Arundinaria japonica (=*Pseudosasa japonica*) **(Ya-dake)**

2–5 m. Hardy. One branch per node. Branching starts in the upper half of the plants leaving long straight culms which were used as arrows (Ya=arrow).

Arundinaria pygmaea (=*Pleioblastus pygmaeus*)
(Oroshima-chiku)

0.3–0.5 m. The smallest bamboo in the world—ideal for bonsai.

BAMBOOS: One of Life's Basics

Bamboos provide food for body and soul, tools for art, for architecture, for agriculture and the tea-ceremony. A plant that supports all Asiatic nations, it reaches its zenith in both utility and culture in Japan. Bamboo groves cover the hillsides, growing wild or farmed for timber.

They are an essential plant in gardens, giving serenity in the country and a glimpse of rural tranquility in city courtyards or as tiny bonsai in a tokonoma. Botanists have decided that their unusual characteristics place them in their own family of about 50 genera and 1,250 species. About 15 types grow commonly in Japan, mainly of the running root (monopodial) division. The clump forming groups (sympodial) are mostly tropical and less hardy. Types range from the tall **madake** and **mōsō**, which together form about 90% of bamboos in Japan, to the dwarf **sasa**. The latter carpet large areas in the north and are used in gardens to make low flat green areas approximating to lawns.

When a bamboo shoot breaks through the ground it is as wide as it will be at its final height and it contains all the joints (nodes) which it will have when it completes its growth. The rapid growth of the shoot (47'6" in 24 hours has been recorded in Kyoto) extends the distance between the joints. Though large bamboos are tree-like, they do not have annual rings and their trunks are known as culms. Canes refers only to dried bam-

SASA

KUMA-ZASA

boos. The shoot, **take-no-ko**, (literally bamboo child) is hugged and protected by sheaths, one attached to each node. These "baby clothes" will be shed as the culm comes to adolescence and will be used as food wrappings and thatch. **Sasa**, however, retains its sheaths all its life. When the shoots are for eating they must be cut whilst young and tender, for with such rapid growth even a day too long can toughen them. The shoots of many types can be eaten but **mōsō** is the most favoured—boiled and then lightly stirfried, or simmered with **shōyu** (Japanese Soy Sauce). This is a spring delicacy enjoyed in April and May.

The timber in a grove is cut selectively when the culms are 3 to 5 years old, and the extensive rhizome network under the soil is constantly regenerating shoots. One rhizome may easily twist its way 100 m. Underground, a bamboo grove is like a convoluted mass of thick plastic netting, preventing mountain sides and river banks from slipping, and being said to be the safest place to be in an earthquake. Above ground the atmosphere is cathedral-like; tranquil, elegant, shaded. The smooth culms support a vault of branches and leaves which rustle with their own gentle music, so sympathetic with the Japanese harmony and instruments of bamboo. The bamboo is truly like the wise man of the Japanese saying who, bending before the winds of life, is never broken.

It is often said that the bamboo flowers only once every 100 years and then dies but this is only partially true. Some species flower almost every year, others have been recorded as flowering irregularly every 50–120 years. These may then die, or the effort of flowering may be so great that the leaves may drop and no food be made or transported to the roots. The plant is thus so much weakened that it may be several years before there is any noticeable growth. The strange fact has been noted that when some types of bamboo flower, all members of that species in the country—and even further afield—also flower within two seasons, amounting to a mass suicide of bamboos. Consequently the "flowering of the bamboo" has come to be a portent of disaster.

TAKE-NO-KO

BARBERRIES

Mahonia *(Hiiragi-nanten)*

Mahonia japonica (Berberidaceae)

Originally from China, but now grows wild on the mountains south of the Kanto Plain and in gardens throughout the world.

The Japanese name is descriptive of the leaves (**hiiragi**= holly) which are made up of holly-like leaves in the arrangement of a **nanten** leaf. They are evergreen on thornless erect stems to 3.5 m and often turn bronze in the winter. The flowers are yellow and fragrant, in pendulous clusters in late winter or early spring, followed by purple-black berries in June. An all-season plant.

The Japanese Barberry *(Megi)*

Berberis thunbergii

A bushy deciduous shrub of 1–2 m growing on hills and mountains all over Japan. Leaves are oval, to 3 cm, and brilliant red in the autumn. Needle-like spines on the branches grow singly. Yellow flowers hanging in 2's and 4's are followed by red, oval, bead-like berries. There are many cultivated forms grown throughout the world, eg *B. thunbergii atropurpurea* and *B. thunbergii atropurpurea nana*.

The Japanese name, which means "eye-tree" was derived from the use of an infusion of the leaves as an eyewash.

HIIRAGI-NANTEN

MEGI

BEANS

Soya Bean *(Daizu)*
Glycine max. (Leguminosae)

Bushy, hairy annual to 60 cm with hairy pods, each containing 1–4 roundish seeds usually whitish but may be yellow, green, red or black. Leaves have 3 large (to 9 cm) leaflets (rarely 5), each broadly ovate. Flowers are inconspicuous clusters of white or violet and are pea shaped.

Very widely cultivated in the warm temperate zones of the world and as a result many cultivars have been developed. Ancestors came from China, Japan and Taiwan but yield outside these areas was only increased when the essential bacterium in the root nodules that enables it to use atmospheric nitrogen (***Rhizobium japonicum***) was identified and spread with it.

The very rich seed up to 20% oil and 45–48% protein, is used for margarine manufacture and in human and animal foods.

Azuki Beans *(Azuki)*
Azuki angularis (=*Vigna angularis, Phaseolus angularis* var. *nipponicus*)

A hairy annual similar to soya bean but a shorter and more compact plant (30–50 cm). Yellow flowers appear in summer in small short-stalked clusters. The long thin pods are yellow blown or blackish when ripe and contain 9–10 small round red beans.

Sekihan, which is made by steaming together rice and **azuki** beans, is served at any celebration and **azuki** cooked with sugar to form **an** is the basis of most Japanese sweetmeats.

DAIZU

AZUKI

Soya Bean Products

Miso. Bean Paste. A very tasty and nutritious seasoning made by crushing soya beans and mixing them with salt and a rice "starter" then fermenting the mix in a wooden tub. It is usually some shade of brown and is a thick paste, though different areas have their own recipes and specialities. ***Miso-shiru*** is soup made with ***miso***, vegetables and often ***tōfu***. It is a basic dish, especially for breakfast.

Shōyu. Soy Sauce. Is made from beans, wheat and salt. Brought from China in the 6th century by Buddhist priests and used as Westerners would use salt, though giving flavour as well. There are different kinds in Japan, mostly lighter than Chinese Soy Sauce.

Tōfu is a highly nutritious, easily digested, protein food made by adding a "starter" to the warm liquid achieved after soaking, grinding and filtering soya beans. When a firm curd is formed, it is pressed to remove surplus moisture, then sliced as required. ***Tōfu*** is always kept under water to keep it fresh. Firmer types can be deep-fried, skewered or grilled. Softer types are used in ***miso-shiru, suki-yaki, shabu-shabu*** etc. In summer, ***tōfu*** is very refreshing chilled on ice and served with ginger, ***myōga***, or young fresh leaves of ***shiso***.

Nattō is fermented soya beans eaten at breakfast.

Above: Making *miso*.
Below left: *Nattō* and *miso-shiru* with *tōfu*.
Below right: Drying soya beans.

BEAUTY BERRY

Callicarpa japonica (Verbenaceae) **(Murasaki-shikibu)**

A deciduous shrub to 2 m with arching stems of sharply elliptical entire toothed, opposite leaves to 5 cm which turn pinkish in autumn. The flowers are inconspicuous. The shrub is well known for its bright purple berries, small, bead-like and almost luminous, which are carried in clusters at regular intervals along the branches, at the nodes, in late summer and autumn. There are also white berried forms.

Murasaki-shikibu is a very evocative name—that of the lady of the Heian period who wrote the famous *Tale of Genji* which tells of court life in the 11th century. (**Murasaki**=Purple.)

BELLFLOWER OR BALLOON FLOWER

Platycodon grandiflorum (=*Campanula grandiflora,*
 C. glauca, Platycodum grandiflorum var. *glaucum*
 etc.) (Campanulaceae) **(Kikyō)**

Native to China and Japan with blue 4–5 cm saucer shaped flowers all summer on perennial plants to 45 cm. It is the inflated 5-angled buds, like traditional paper balloons, which give the flower its name and make it unique. The leaves are oval-pointed, with almost no leaf stalk, 4–7 cm long and light green. Cultivars include double and white flowered forms and forms in which the flowers retain the balloon shape of the buds.

A traditional and popular decorative motif. One of the Seven Flowers of Autumn.

MURASAKI-SHIKIBU

KIKYŌ

BUCKWHEAT

Fagopyrum esculentum   (Polygonaceae)   **(Soba)**

An annual crop-plant with floppy reddish stems and long gently triangular leaves, the upper ones clasping the stem. The flowers are in short, dense, long-stalked clusters, white in Japan, pink in other countries. The seeds are sharply 3-sided about 5 mm across. Hokkaido is the main production area.

Soba (buckwheat noodles) is a staple and traditional food of the Japanese. It is particularly a lunch time food but is taken as a snack at anytime of the day or night. Dried **soba** seeds are infused to make a nutty drink. **Soba** chaff is used as stuffing for pillows.

BURDOCK

Arctium lappa   (Compositae)   **(Gobō)**

A coarse herbaceous biennial of up to 1.5 m that grows widely in Europe and northern Asia but developed and cultivated only in Japan. The leaves are oval-heart shaped to 40 cm, the basal ones with furrowed leaf-stalks. The purple flowers appear in summer and are the familiar sticky-bobs of country children. Botanically they are a sphere of hooked involucral bracts almost hiding an inflorescence of purple flowers made up of disc florets (tubular) only.

Originally both seeds and roots were used medicinally. Now it is the long (to 70 cm), brown-skinned roots that are used as a vegetable. Thin strips are sautéed to make a popular side-dish (**kinpira gobō**) or to give flavour to winter soups and stews.

SOBA

GOBO

BUSH CLOVER

Lespedeza bicolor (Leguminosae) *(Hagi)*

Deciduous sub-shrub, sometimes reaching 2.5 to 3 m. Arching branches grow in clusters from a single rootstock each spring and are clothed with trifoliate leaves, 7–8 cm. Pink pea flowers come in panicles at the ends of the branches in late summer. In gardens the plants are cut right to the ground after flowering. There are many species of *Lespedeza* growing wild on mountains and hillsides in Japan, and in gardens throughout the world.

In Japan they are typical of the way that the garden achieves harmony with the landscape by bringing wild plants into important situations and revering them. The bush clover, the only plant in the courtyard of the Seiryo-den of the Kyoto Gosho Palace is a supreme example of this harmony and reverence for nature within the exuberances of court life. The arching branches and rustic charm of the plant appeal to the Japanese and, in the blazing heat of late summer, promise cooler, more serene autumn days. One of the Seven Flowers of Autumn. (Appendix 2) See them in Sendai Hagi Park, Miyagi Prefecture.

BUTTERBUR

Japanese Butterbur, Bog Rhubarb *(Fuki)*
Petasites japonicus and *Petasites japonicus* var. *giganteus*.
(Compositae)

A perennial of woods and damp places with tough invasive rhizomes. On female plants, the flower stem appears in early spring before the leaves, at first a greenish yellow knob breaking through the ground, then elongating to a stout stem with many scale leaves, topped by an ovoid conical head of many button shaped inflorescences of rayless flowers. Eaten in tempura before it elongates too much. The round leaves follow quickly to become very large, and with long, edible leaf stalks.

Petasites japonicus is the smaller and more common variety, with flower stalks between 20–80 cm long, and leaves 20–70 cm. It tastes better than the massive *P. japonicus* var. *giganteus* found in the Akita region of N.W. Honshu, which can grow to over 2 m with leaves of 1.5 m across—useful umbrellas in a shower!

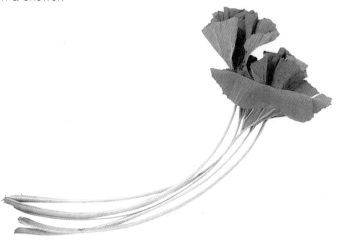

FUKI

FUKI-NO-TŌ

CAMELLIAS

Camellia japonica   (Theaceae) ***(Tsubaki)***

Grows wild in the forests of the southern mountains, where it becomes a straggly tree to 10 m with smooth grey bark. The evergreen leaves are oval, glossy, dark, with finely toothed margins. The single red flowers of the type plant come Nov.~Apr. and have a beautiful central boss of golden stamens. **Tsubaki** has been cultivated for centuries and there are hundreds of cultivars—singles, doubles, striped and blotched; white through pink to red. The first main shipment to Europe was in 1792 through the East India Company, though they had been known earlier in the century.

Tsubaki oil is obtained from the seeds and was used as hair oil until the 1930's to maintain men's top knots and the high coiffure of Japanese ladies. **Tsubaki** oil is also used for deep-frying tempura.

Characteristically, the flower heads of **Tsubaki** drop off whole, not petal by petal. This made the samurai consider them an unlucky emblem, seeing in them too much resemblance to human heads falling.

Camellia sasanka   ***(Sazanka)***

Very similar to *C. japonica* but flowers in autumn and winter, and is often seen as a hedge or clipped into a cone for tubs and pots. The flowers are slightly fragrant, single or double, white, pink or red. They can be distinguished from **Tsubaki** by the way the flowers fall petal by petal, and by the hairs on the young shoots.

Tea

Camellia thea   ***(see under "Tea")***
page 196

TSUBAKI

SAZANKA

CAMPHOR TREE

Cinnamomum camphora (Lauraceae) ***(Kusu-no-ki)***

A tree of the warmer areas from Tokyo southwards. Although it is an evergreen, it is the clouds of fresh green leaves billowing in the upper tree canopy that make it noticeable in May. Later the leaves are dark green and leathery, slender and pointed, with a prominent mid-rib and two clear lateral veins. Flowers in May/June are small, green and inconspicuous. Camphor is found in almost every part of the tree and can be sensed on a sunny, humid day in late spring. It is a very long-lived tree, widely planted in parks and shrines. Some may reach 55 m in height and over 7 m across. Consequently there are many legends about Camphor trees, and many trees that are sacred and venerated.

KUSU-NO-KI

KUSU-NO-KI

CEDAR

Japanese Cedar *(Sugi)*

Cryptomeria japonica (Taxodiaceae)

A very tall, sharply conical, evergreen conifer, 30–40 m, some more than 60 m. A very important tree in the landscape, forestry, traditions and legends of Japan. The foliage is sharply awl-shaped on long slender branchlets and is much more feathery when young. Some cultivars stay dwarf and keep their juvenile foliage throughout their lives (eg *C. japonica elegans, C. j. compacta* and *Bandai-sugi, Jindai-sugi*, etc.). The globular female cones develop in the first year at the ends of the shoots but do not become brown until the second autumn when they dry out and release the seeds. The bark is reddish brown and peels away in thin vertical strips.

The tall straight trunks make ideal flag and telegraph poles, and Cryptomerias can still be seen marking boundaries. In the Kitayama Forestry Reserve, NNW of Kyoto, most beautiful straight trunks are produced by a kind of coppicing. Trimmed and polished, they are used for tokonomas and for rafters. The wood splits easily so that, even before sophisticated implements were available, it could be made into planks which were widely used for making boats, houses, boxes, sake barrels and chopsticks. The bark is also sliced off to make shingles for shrine roofs. A green ball of *Cryptomeria* foliage is hung outside sake breweries when the sake is finished in March. By the time the ball turns brown in summer the sake is ready for sale.

The magnificent 20 mile avenue of Japanese Cedars leading to the Tōshōgu Shrine in Nikkō is a lasting memorial to Masatsuna Matsudaira who rebuilt the shrine in 1625. At the time he was ridiculed by the other Lords for giving this avenue when their more appropriate gifts were lanterns and other ornaments. *Cryptomerias* are often seen in shrine precincts with ropes round their trunks to mark them as sacred.

SUGI

CHERRY BLOSSOMS

Prunus spp. (Rosaceae) ***(Sakura)***
The Mountain Cherry ***(Yama-zakura)***
Prunus jamasakura (=*P. serrulata*)

Believed to have originated on Mount Yoshino near Nara. Five notched petals. Leaves soft red-brown, appearing with the flowers. Flowers white to pink and even darker. Among the cultivars are double flowered varieties.

The **Ō-yama-zakura** is a hardier type with deeper flowers, and that found high on the mountains of Nagano is quite dark red. It is the most highly regarded cherry by the Japanese and is the prototype of most of the others.

Different authorities name anything from 30 different kinds to over 400, and parentage is consequently confusing. **Yama-zakura** flower in the hilly areas almost one week after the **Somei-yoshino** in towns. See them in early-mid April on Mount Yoshino, S.E. of Osaka, at Kiyomizudera Temple and at Arashiyama in Kyoto, and in the hills of Kamakura.

The Tokyo Cherry ***(Somei-yoshino)***
Prunus x yedoensis

This cherry has branches packed with single flowers which appear before the leaves, shell pink in bud, fading to almost white. It is quite a tough little tree, maturing in under 20 years and having a short life compared to many cherry trees. The tree is a hybrid which appeared just over 100 years ago reputedly from a single plant in a nursery in Somei, in Sugamo (Tokyo). The first known batch of them was planted in 1872 when the Imperial Museum was being built in the area which is now the Imperial Hotel in Tokyo.

Most plantings in towns and cities are now of the **Somei-yoshino**, so you will see them everywhere.

YAMA-ZAKURA

SOMEI-YOSHINO

The Weeping Cherry
Prunus pendula (=*P. itosakura*)   ***(Shidare-zakura)***

May be white or pink, and is very long lived. One is known to be more than 1000 years old, and another is 10 m round the trunk. There are famous ones at Daigo-ji and Maruyama Park in Kyoto, and a fantastic arbour of them at the Heian Shrine in Kyoto. There is also an aged tree at Miharu in Fukushima Prefecture. Some weeping cherries have originated from other species, e.g. **Shidare-yamazakura**.

The Fuji Cherry   ***(Mame-zakura)***
Prunus incisa

Usually 3–5 m. Its restricted and slow growth makes it the most popular cherry for bonsai. The profusion of flowers which cover the branches are small and white, though pink in bud. Leaves are small and much toothed, colouring vividly in autumn. **Mame-zakura** grows wild on Mount Fuji.

Double Cherries   ***(Sato-zakura)***
Garden Hybrids of obscure parentage

These cherries have been cultivated, hybridised and grafted for over 1,000 years. **Sato**=village, so the **Sato-zakura** are the cultivated cherries, as opposed to the **yama-zakura** from the hills. The **Yae-zakura** are double and are the emblem of Nara—where you can see them. All flower later than the **Somei-yoshino** and come in all shades of white and pink, and even yellow. They have very full flowers with up to 30 petals.

  Sato-zakura were collected, hybridized and widely planted by the Tokugawas. They spread throughout the world in Meiji times and their popularity still increases. The most comprehensive collection is in Southern Hokkaido where Matsumae Park

SHIDARE-ZAKURA, BENI-SHIDARE

YAE-ZAKURA

boasts 8,000 trees in 250 varieties. The 3,000 trees planted along the Eastern bank of the Arakawa river in Tokyo in Meiji times have all been lost but replanting has started near Nishi-arai bridge and it is hoped to build up the collection to 1,000 trees. Along with all this interest, the taxonomy of **Sato-zakura**—indeed of all Japanese Cherries—is under constant revision.

Fugenzō is a beautiful, internationally known, pink variety which was first recorded in Japan in 1555. **Kanzan** spread widely in Europe in the Meiji period (1868–1911). **Ukon** has large semi-double flowers of a greenish pale primrose, and **Gyoikō** is another well known greenish one. **Kiku-zakura** means 'Chrysanthemum flowered cherry' and has flowers like pink pompom chrysanthemums. In addition to these there are well over 100 varieties known and listed.

View **Sato-zakura** in Shinjuku-gyoen in Tokyo in the second half of April, home to the Emperor's cherry viewing party. In 1912 the people of Tokyo gave 1,000 **Sato-zakura** and 1,000 **Somei-yoshino** to the people of Washington, USA, where they are planted around the Tidal Basin.

CHERRIES: The Soul of Japan

The natural beauty of the individual flowers, the fragility of the blossoms under blue spring skies, threatened by showers and gusts of wind, have given cherries a special place in the hearts of the Japanese from the latter part of the 8th century, when the aristocracy first began to recognise their beauty. The Emperor Jito (690–697) is known to have taken parties to view the blossoms on Mount Yoshino. Viewing Parties—**hanami**—became more and more popular and elegant as the courts became more sophisticated. Nowadays they are enjoyed by absolutely everyone. One could not now pass on the advice of Florence Cane, writing in 1908, that "the foreigner wishing to enjoy... peace for his viewing, will do well to spend a few hours in undisturbed enjoyment of the more dignified setting of Ueno Park"*. The cherries at Ueno are relics of the many groups planted by

SATO-ZAKURA

YAMA-ZAKURA

the Tokugawas in the Tokyo area after 1603, and are certainly incredibly beautiful, but the park is now so crowded that it needs a vigorous imagination to see the elegant ladies in their special silk **hanami** kimonos, hear the music of the **shamisen** and participate in the writing of succinct **haiku** verses. The **sake** is still there, the music is still there—though now mostly canned 'backing' music. Above all, the "spirit of sakura" is still there—and being enjoyed by millions.

"The soul of **sakura** is the soul of **bushido** (chivalry), and the heart of **bushido** is the heart of Japan"—according to the samurai, who were expected to die willingly, like cherry blossoms swept away by the sudden breezes, whenever their time came and however short their moment of glory had been.

One way to retain the fleeting moment of the cherry blossom is to salt them. These salted flowers are then infused to make the tea which is given to a bride before she goes to her wedding ceremony. The flower opens out in the water and seems to blush, in sympathy with the bride. Leaves are also salted and used to wrap the sweet bean-paste cakes (**sakura-mochi**) which are popular at cherry blossom time.

The wood of the cherry tree is very hard and makes long-lasting blocks for wood-block printing. A veneer of cherry bark is a very beautiful finish for small furnishings, tea caddies, etc.

Find out about the northwards progress of the "cherry blossom front" from the TV or *the Japan Times* or by asking at your hotel desk. Through April into May depending on area.

See Appendix III for classification of cherry blossoms.

*The Flowers and Gardens of Japan, p. 139 by Florence du Cane, 1908.

CHRYSANTHEMUM

Edible Chrysanthemums *(Shun-giku)*
Chrysanthemum coronarium var. *spatiosum* (Compositae)

Orignally from Southern Europe, developed in China and grown in Japan as a green vegetable for spring or autumn sowings. Though cut for use when about 15 cm high, it grows naturally to 30–60 cm. It is much branched, hairless and with leaves twice-cut into quite narrow segments. Single yellow or white, rayed, composite flowers, 3–6 cm. Stems soft and succulent.

The leaves are boiled and served simply with soy-sauce or with sesame and often used in **suki-yaki** and other cook-at-the-table dishes.

Some cultivars of *Chrysanthemum morifolium* below are also grown as a vegetable but it is the petals which are eaten blanched and dressed with vinegar.

Florists Chrysanthemums *(Kiku)*

Developed over 1,000 years, probably from *C. indicum, C. makinoi*, and other Chinese and Japanese species. Nowadays generally grouped together as *C. morifolium* and then classified into groups according to flower size and characteristics. They are grown worldwide, but surely reach perfection in Japan, the land of the chrysanthemum. The 16–petalled chrysanthemum has been the official crest of the Imperial Family since Meiji days and used by them as a personal motif and decoration long before that.

The cultivation as pot plants is more apropriate to Japanese spaces, inclinations and natural skills than is the cultivation of massed displays. The meticulous feeding, watering, training and tying are almost ritualistic. The care is fastidious, the beauty is disciplined, the goal is perfection.

Exhibitions and shows are held in most towns about the first week of November, with special ones in Tokyo at Shinjuku-

SHUNGIKU

BONYO-GIKU

gyoen, Yasukuni Shrine and Hibiya Park, and near Osaka at Hirakata Park.

Exhibition Chrysanthemums   *(Kanshō-giku, Bonyō-giku)*

This is the most familiar group to Western growers with three stems per plant, one bloom per stem. "**Large Flowered**" have flowers more than 6″ diameter, "**Medium Flowered**" 2–6″, "**Small Flowered**" less than 2″. Many classifications: anemone flowered, incurved, reflex, tubular, spoons, brushes etc. Rich composts may contain rice bran and rice-hull charcoal. Feeding, watering and training will be constant. Three bamboo stakes are tied to each of three holes found just below the rim of the pot and the massive flowers supported and controlled by a circular wire arrangement (***rindai***) which can be moved up and down the stem as required.

Bonsai Chrysanthemums

A comparative newcomer to bonsai work and particularly well suited, with pliant wood and many flowers on short stems. Special cultivars with short internodes have been developed. An aged specimen can be achieved in 12–14 months, so mistakes can be rectified on next years attempts. From September cuttings special training and pinching goes on all year so that simultaneous blooming over the whole of the plant is achieved in late October and early November.

Cascades   *(Kengai)*

A solid mass of tiny flowers, in the shape of a waterfall or an apron, that may vary in size from a miniature bonsai to a massive exhibit in a show, an hotel foyer, or other special place. The young plant is potted at an angle of 45°, staked, and as it grows it

BONSAI

CASCADES

is gradually lowered and trained to a wooden V-shaped frame with crossbars and later to a wire frame. Constant pinching and inter-weaving of shoots gives the solid mass and simultaneous blooming.

The understatement of all time must be that of the Japanese grower who said that these masterpieces "are no trouble to achieve ... only needing good stock, hard work and patience"!

Thousand Bloom Styles

Massive geometric displays of 300–500 medium or large flowered blooms evenly spread on one plant. Widths of 2.5 m and overall heights of 2 m are not uncommon—grown as semi-hemispheres or pyramids. Varieties with long internodes, plenty of branches and of a fairly regular flower form are necessary. Tubular and spider forms have petals that would tangle together.

Chrysanthemum Dolls *(Kiku-ningyō)* *(Chū-giku)*

A craft that started in the early 19th century and which led to public displays that were prestigious and popular in north-eastern Edo (Tokyo) where many of the city's gardeners had their homes. After this their popularity waned, but now there is a revival.

It takes 4 years to make the larger models. For the smaller ones, young plants are lifted from the ground and placed inside body frames made of straw and bamboo and their roots embedded in sphagnum moss. By flowering time the frame is completely hidden.

In some places there are elaborate stagings of plays, or of moments in history, with both set and characters worked out in chrysanthemums. There are special exhibitions at Hirakata Park near Osaka, and at Nihonmatsu in Fukushima Prefecture.

KIKU-NINGYŌ

CITRUS FRUITS

(Rutaceae)

Japan produces most of its own citrus fruits, mainly Satsuma types, or **mikan** as they are known in Japan. The main production areas are in the warmer south, especially along the Pacific coasts from Atami to Kyushu. Kyushu is warm enough for lemons but the high summer humidity makes pest and disease control a problem.

In the shops you will find **Mikan** from June to March, **Hassaku** in March/April, and **Natsu-mikan** from the end of March through to May.

Satsuma (Mandarin) *(Mikan)*
Citrus unshiu

A variety developed in Japan, and now widely grown in southern and central parts. A tree to 3 m with no thorns. The fruits are flattened, 5–10 cm diameter. They have all the virtues to make them the most important of all the Japanese **mikan**; sweet and juicy, very easily peeled with thin skins and usually no pips. It is hardy and stores well. Early maturing varieties are ready from early September, later ones in November.

Summer Orange *(Natsu-mikan)*
C. natsudaidai

A larger fruit, about 10 cm deep, 12–18 cm across with very thick, rough skin that needs a knife to it. It is edible but sour. The flowering period is in late spring and although the fruit is ripe in the winter, it is not edible until the following late spring or early summer. This overwintering of the fruit means that it needs warmer winters than does *C. unshiu*, or the protection of Tokyo's micro-climate.

UNSHŪ-MIKAN

Hassaku *(Hassaku)*
C. hassaku

This is a hybrid **mikan** which ripens in March/April and grows in similar areas to *C. natsudaidai*.

Pummelo *(Buntan, Zabon)*
C. grandis

An enormous round yellow fruit of 17 cm diameter and over with very thick skin and firm flesh which is meaty and sweet rather than juicy. Originally it came from East India and Burma and is now grown in Kyūshū. It ripens in January/February.

Kinkan *(Maru-kinkan)*
Fortunella japonica

A bushy evergreen to 2 m found in warmer parts, fruiting in November/December, bearing small round 'oranges' only about 2.5 cm in diameter. The skin is thin and waxy and the fruit very sour with lots of pips. Cooked slowly in a heavy sugar syrup, the fruit is important as a New Year food and the resulting syrup as a remedy for coughs.

Kumquat *(Naga-kinkan)*
F. margarita

Similar to *F. japonica* but with somewhat sweeter oval fruits about 4 cm long with thin edible skins. They can be eaten raw or cooked and preserved, used in sweet savoury dishes and made into wine or marmalade.

BUNTAN

MARU-KINKAN

Yuzu *(Yuzu)*
Citrus junos

A small evergreen tree to 4 m, also often grown as a pot plant. The small (4–7 cm), rough-skinned fruits ripen to yellow in December. The juice, and more particularly the skin, is indispensible to the New Year cuisine, where fine strips add fragrance to soups and other dishes.

Japanese Bitter Orange or Trifoliate Orange *(Karatachi)*
Poncirus trifoliata   (=*Aegle separia*)

Usually seen as a spiny impenetrable hedge. Though it loses its leaves in winter it remains colourful because of its green stems and spines. The leaves are in threes. The large 5-petalled, white flowers are about 2 cm across and appear in the spring as the leaves are about to emerge. They are followed by little oranges, 5 cm, that are round and bitter, green, ripening to yellow, fragrant rather than edible.

YUZU

KARATACHI

CYCADS

Cycas revoluta   (Cycadaceae)   ***(Sotetsu)***

Slow growing palm-like plants. The single trunk with a cluster of leathery fern-like fronds on top can eventually reach 5 m. Male and female cones are produced on separate plants in the centre of the leaf cluster, the male cones growing to 60 cm and shedding masses of pollen.

Botanically, they are very ancient and primitive. Architecturally, they give a strong line, with added impact when grown in the usual grouping of 3, 5 or 7. Horticulturally, they are only reliably hardy in the warmer parts, though with traditional Japanese ingenuity and neatness, the trunks are bound in winter with rice straw, tied with immaculate knots, and capped to keep out the cold.

There is a famous group on Cycas Hillock at Katsura villa in Kyoto.

SOTETSU

CYPRESSES

White Cedar *(Hi-no-ki)*

Chamaecyparis obtusa (Cupressaceae)

A native forest-forming species to 36 m and more. A broadly conical, rather open, evergreen with an egg shaped top. Blunt-edged leaves in flattened sprays with a distinct bluish line underneath. Cones small (10 mm) with 8 scales with small points.

Widely grown for its top quality, valuable, building timber, especially on the Pacific Ocean side. On the Sea of Japan side it can suffer damage from heavy snowfalls. There are many cultivars and in spite of its natural size it makes a good bonsai subject. Other cultivars are good hedging plants.

It was such valuable timber in the Kiso Valley that any peasant who felled one of the protected White Cedars was liable to execution!* Its particular value is that it does not split or warp. Hence its use as a high quality building wood and for measures for dry goods (beans and rice etc.). A Noh stage is constructed entirely of White Cedar. You may also enjoy the scent of the wood, either in a traditional wooden bath tub or when drinking sake from the smallest of the square measuring boxes.

Sawara Cypress *(Sawara)*

C. pisifera

Distinguished from *C. obtusa* by its sharply conical top and the sharply pointed adult leaves. The lower branches tend to be shed when old, revealing a distinct and typical bole. Cones are about 15 mm, with 10 pointed scales. There are many cultivars. The juvenile leaved forms, eg. *C. pisifera plumosa*, remain feathery when mature and were originally listed as *Retinospora*.

Yoake Mae (Before the Dawn) by SHIMAZAKI Tōson

HI-NO-KI

SAWARA   HI-NO-KI

DAIKON

Japanese White Radish *(Daikon)*
Raphanus sativus (Cruciferae)

A rough, hairy biennial, grown as an annual, reaching 1 m in height. Flowers may be white to pinkish-lilac, 1–2 cm. The lower leaves are pinnately lobed with a large rounded terminal lobe. The swollen white root is usually cylindrical and up to 70 cm long but is spherical in some varieties.

It is the most commonly eaten vegetable in Japan and the most popular pickle. Grated, it forms the Fuji shaped pile on the plate that accompanies your tempura. Pickled, it becomes yellow and turns up in **sushi**, as well as on your breakfast tray—its high diastase content helps the digestion of rice. **Daikon** is also used in soups and stews.

DAPHNE

Sweet Daphne *(Jinchōge)*
Daphne odora (Thymelaeaceae)

Intensely fragrant flowers in late winter herald the spring. Originally from China, it is now widespread and is often grown as an evergreen hedge. Left unclipped it will reach 1.5 m. The leaves are leathery and broadly lance-shaped, those at the ends of the branches surrounding a cluster of flowers. The flowers may be white, red or pink—or even white on the insides of the petals and red or pink on the outsides. Most are male and do not fruit.

It is said that **Jinchōge** should only be transplanted when it is in flower, as the cool temperatures at this time allow any of the very vulnerable broken roots to heal, rather than rot. It is propogated from cuttings.

DAIKON

JINCHŌGE

DAWN REDWOOD

The Fossil Tree *(Akebono-sugi)*
Metasequoia glyptostroboides (Taxodiaceae)

A deciduous conifer that grows to about 35 m in the wild.

The narrow leaves, each about 2 cm, are arranged oppositely and are a fresh green when they appear in the spring as tufts. These quickly elongate into branchlets which turn mellow bronze and yellow before falling in the autumn. Cones are round, about 2 cm, on long stalks. Bark is dark grey/brown and shaggy.

A plant that was thought to be extinct until 1941 when it was discovered growing in China. Before then it was known only as a fossil in tertiary rocks. The only known living member of the genus. It is interesting to see a glade of two plants of great interest to botanical historians—Metasequoia and Ginkgo—growing together in the Koishikawa Botanic Garden in Tokyo.

AKEBONO-SUGI

DAY LILY

Hemerocallis fulva   (Liliaceae)   **(Yabu-kanzō)**

A rhizomatous, clump-forming, perennial, 70–100 cm high with narrow, bright green leaves. Trumpet shaped orange-yellow flowers, about 10 cm each last one day but a successive flowering on the strong stalks gives them a long period from early to mid-summer. Grows abundantly in many situations; on the edges of paddy fields and woods, or mountain sides, in gardens, in sun or shade, in moist or dry.

The leaves can be eaten when young. So can the flowers, though they look better than they taste. The roots were at one time used as a sweetener.

The golden yellow form, the **Nikkō-kisuge**, grows in profusion at Nikkō.

DIANTHUS

Fringed Pink *(Nadeshiko)*
Dianthus superbus L. (Caryophyllaceae)
(Kawara-nadeshiko)

A widespread hardy perennial growing to 50 cm. **Nadeshiko** is a general term for mainly native *Dianthus* groups. The leaves are linear and opposite with prominent nodes. the flowers are solitary or few to a stem with long calyces. They appear in various shades of pink from mid-summer onwards. The intense fringing of the petals camouflages the fact that there are only five of them. The delicate beauty of the flowers belies the hardiness and strength of the plant. This has given it a long tradition in literature and a place as the symbol of the ideal woman; a combination of strength and grace. It is cherished as one of the Seven Flowers of Autumn. (Appendix 2)

DOG'S TOOTH VIOLET

Erythronium japonicum (Liliaceae) *(Katakuri)*

One of the most declicate and graceful species of a genus spread throughout the world. Found growing in light woodland on low mountains in north-eastern regions, flowering in April with six reflexed purplish-pink petals and long black pendulous stamens. Two broadly lanceolate stalked leaves come out from the bulb which yields a starch once used widely as a thickening agent in Japanese cooking. Real **katakuri-ko** is now quite hard to obtain and has been replaced by potato starch. It seems incredible that such a dainty dancing flower and its attendant leaves can be eaten but it occasionally appears as **tempura**.

KAWARA-NADESHIKO

KATAKURI

DOGWOODS

Japanese Strawberry Tree *(Yamabōshi)*
Cornus kousa (Cornaceae)

Native to Japan and Central China, **Cornus kousa** is the most showy form. Spreading, deciduous shrubs/small trees 5–10 m, covered along the tops of the branches in May with masses of upright spreading flowers. The four creamy white pointed petals are really bracts, and the real petals would be taken, at first glance, to be stamens. The leaves are oval pointed, to 10 cm, opposite, and develop beautiful bronze and crimson autumn tints. The fruits look somewhat like strawberries.

Flowering Dogwood *(Hana-mizuki)*
Cornus florida

Native to Eastern U.S.A., and very similar to **Cornus kousa** but the bracts of the flowers are rounded. Flowers are typically white but there are many pink cultivars. The branches are in layers and are covered with upright stalked flowers in May.

There are fine displays of Dogwoods in Tokyo in Koishikawa Botanic Gardens, Shinjuku-gyoen and Hibiya Park. The latter were a gift from the people of Washington in 1915 in return for the gift of cherries (see page 54).

YAMABŌSHI

HANA-MIZUKI

ENKIANTHUS

Red Veined Enkianthus *(Sarasa-dōdan)*

Enkianthus campanulatus (=*Andromeda campanulata*) (Ericaceae)

White Enkianthus *(Dōdan-tsutsuji)*

Enkianthus perulatus (=*Andromeda perulata*)

Both are deciduous shrubs to 3 m with fiery, purple-red leaves in autumn. Occasionally found wild in the mountains west of Izu and frequently grown in gardens as a shrub and a hedge. The branches are in layers, and the narrow pointed leaves, to 6 cm long, are in whorls. In spring the lily of the valley shaped flowers hang down in bunches, those of *E. campanulatus* being pinkish and bell shaped, whereas those of *E. perulatus* are pearly white with constricted openings.

SARASA-DŌDAN

DŌDAN-TSUTSUJI

EURYA

Cleyera japonica (=*Sakakia ochnacea,* **(Sakaki)**
Eurya ochnacea) (Theaceae)

An evergreen tree, 8–10 cm, growing from the Kanto area southwards and usually planted in shrines as it is the sacred tree of Shintō. The leaves are thick, shiny and alternate, up to 8 cm long, narrowly ovate with pointed ends. The bud at the top of each shoot is bent into a bow shape like a claw or beak. In June-July the tree has small creamy-white, bell shaped flowers that hang in short stalked clusters. The fruits are little berries that turn from green to black over the winter.

It is traditionally believed to be the first tree that grew out of the chaos of creation, and a pair of **Sakaki** are often planted in the front of a shrine. The branches are offered at altars, at ceremonies, and are also sold at florists in small bunches for home altars.

Eurya japonica **(Hisakaki)**

Grows both wild and cultivated in gardens in colder regions. Used in Shinto rites. An evergreen shrub 2–5 m high with yellowish-white bell-like flowers in March-April and leaves similar to **Sakaki** but more toothed. The black berries may still be on the branches when the flowers appear.

The Japanese name means "small sakaki".

SAKAKI

HISAKAKI

FATSIA

Japanese Fatsia, False Castor Oil Plant　　　　*(Yatsude)*
Fatsia japonica　(=*Aralia japonica, Aralia sieboldii*)
　(Araliaceae)

Medium sized shrub to 2.5 m, a very architectural plant and often used in courtyard gardens since it enjoys the sheltered semi-shade and will stand pollution. Deeply lobed evergreen leaves, dark green, shiny and leathery, 30 cm or more, on long stalks. Flowers add to the architectural effect in winter. Creamy white and individually small, they are collected together into balls on long, stiff, upright, branched inflorescences. The female flowers develop into black berries in April-May.

GARDENIA

Gardenia jasminoides　(Rubiaceae)　　　　*(Kuchinashi)*

An evergreen shrub, 2–3 m, growing wild at the edge of woods in the warmer regions. Usually seen growing in gardens—often as a hedge where the climate is sufficiently warm and humid. The leaves are dark green and leathery, similar to those of Camellia. The flowers are waxy and white, produced mainly in June and July, 7–8 cm across and fragrant. They too are similar to camellias, especially the double garden hybrids. The strangely ribbed red fruit, about 3 cm across with 6 long horns at its apex, contains many small seeds and gives a dye which is used as a yellow food colouring and a medicine. The fruit does not open, hence its Japanese name which means "no mouth".

YATSUDE

KUCHINASHI

GINGER

Zingiber officinale   (Zingiberaceae)   ***(Shōga)***

Herbaceous perennial 30–50 cm. Grass-like leafy stems arise from fleshy rhizomes which are the source of the spice. The greenish-yellow flowers have 3-lobed purple lips and are in dense spikes. It is one of the oldest spices known and is mentioned in works by Confucius. It needs a warm, moist climate. In spring the shoots with the new rhizomes are served as an accompaniment to fish. The new ginger of summer is fresh and juicy. That of the autumn harvest is used all winter. It is an essential of Japanese cuisine, used in many ways, especially in the basic combination of soy sauce, ginger and sake. A traditional cold cure is **shōga-yu**—made by pouring hot water over grated ginger.

Zingiber mioga   ***(Myōga)***

A native plant of 40–100 cm in height, found in shaded areas in the mountains but also cultivated and used as a vegetable and herb. Strange little yellowish-white hooded flowers with many layers of pink sheaths around them push up through the ground in early summer. The young flowering shoots are edible when about 3–4 cm high. Legend says that if you eat it you will become forgetful.

SHŌGA

MYŌGA

GINKGO

Maidenhair Tree *(Ichō)*

Ginkgo biloba (Ginkgoaceae)

A deciduous tree that grows slowly to 30 m, is conical when young, fan shaped and very graceful later—unless pollarded, as so often happens in city streets. Leaves (5–8 cm) are fan-shaped, more or less cleft into 2 lobes and similar to maidenhair fern. They turn glowing yellow in the autumn. The bark becomes characteristically ridged, cracked and fluted with age. Fruits of female trees are the shape and size of small plums. The edible nut (**ginnan**) is covered with slimy evil-smelling flesh which is removed by soaking the fruits in water, then washing and drying the white shells. These are then cracked and removed to reveal the edible green kernel which may be roasted or used in soups and in **chawan-mushi**. It is rich in vitamin B.

Fossils show that Ginkgo trees were widespread in prehistoric times, but this species, native to China, is the only known representative nowadays. It is of particular botanical interest in Japan because it was a scientist working at Tokyo Imperial University who, in 1896, was the first to discover the spermatozoids of Ginkgo, and so unravelled its reproductive cycle.

There are old and respected ginkgo trees in many Shinto shrines (marked by their encircling ropes and/or white papers) held to be the residence of sacred spirits. The ancient ginkgo still standing by the steps of Tsurugaoka Hachiman Shrine in Kamakura has a special place in history: it provided cover for Shōgun Sanetomo's assassin in 1219.

GINNAN NUT

ICHŌ

GRASSES

Silver Grasses, Japanese Plume Grass *(Susuki)*
Miscanthus sinensis (=*Eulalia japonica*) (Graminae)

A stout, clump-forming perennial, to 2 m high when in flower and with smooth stems. The leaves are slightly toothed, rough and linear, 100–150 cm long, green with white midrib. Flowers are in yellowish-brown tassels that are 20–30 cm long in late summer.

An essential part of the Japanese landscape, growing almost everywhere, **Susuki** becomes especially noticeable in the autumn when the seeds ripen and it becomes silky, light and bright, giving a shimmering sheen to whole mountain sides as well as to odd corners. Many cultivars have been developed: **Hoso susuki**, with narrow leaves, **Takanoha susuki** (*M. sinensis zebrinus*) with yellow horizontal bands on the leaves and varieties developed for bonsai work.

The reeds were at one time used for thatching roofs. It is one of the seven flowers of autumn.

Miscanthus sacchariflorus *(Ogi)*

This is generally larger than *M. sinensis* (to 2.5 m) and is found mostly in swampy places. It is rhizomatous and so is spreading rather than clump forming. The leaves are long and narrow with a pale midrib. They all come from the stem and are gracefully drooping. The flowers are longer and thicker than those of **Susuki**.

SUSUKI

OGI

LAWNS AND GOLF COURSES

Winters in Japan are cold and dry, summers are hot and wet, especially in the Kanto and Kansai. The only grasses that will survive these conditions go brown in winter. *Zoysia japonica* is a native grass of upland areas and roadsides which is often used for lawns. It is in fact sometimes known as 'Japanese Lawn Grass'. It is a very hard, coarse grass, which, even in the rainy season, can be left a week between cuttings. It is usually established vegetatively. Fine leaved, evergreen grasses are reserved for the greens of golf courses, where they can have the necessary attention lavished upon them. This includes almost impossibly frequent cutting in the rainy season and watering in the long dry spells. In winter they stand out as green oases in a desert of brown *Zoysia*.

Under these circumstances, it becomes apparent why the Japanese have used so much moss for 'lawns' in their gardens—mosses delight in the humidity of summer and they remain green throughout the winter.

HEAVENLY BAMBOO

Nandina domestica   (Berberidaceae)   ***(Nanten)***

A stiff cane, occasionally reaching 2 m, topped by graceful evergreen leaves which resemble *Mahonia* (to which it is related), but the leaflets (3 or 5) are narrower, less toothed and more delicate. They are flushed red in spring and purplish red in autumn. The small white flowers in mid-summer come on erect stalked panicles. They give way to sealing wax-red berries (about 5 mm across) which brighten up winter days. This lamp-lighting effect is particularly noticeable as the **Nanten** are traditionally planted at the sides of front doorways or backyards.

It grows wild in woods in the southern part of Japan, is cultivated widely throughout, and many cultivars have been developed with variations in size, leaf colours, and colour of the berries. The wood is very close grained, aromatic, and was traditionally used for chopsticks.

NANTEN

NANTEN

HYDRANGEA

Hydrangea macrophylla (Saxifragaceae) ***(Ajisai)***

The wild prototype grows in light woodlands throughout Honshu, especially in Kanto and on the Izu coasts. A deciduous shrub to 2 m with opposite, entire, toothed leaves. Flowering from early summer, with flat heads of large, blue or pink sterile florets at the outside of the flower head, and small fertile ones in the centre. ***Ajisai*** has given rise to many cultivars, notably the mop-headed Hortensia group with round heads of almost all sterile florets and the flat headed lace-cap varieties. In Japan, the lace caps are called ***gaku-ajisai***—***gaku*** meaning a picture frame. See them at Meigetsu-in and Mandara-dō-ato in Kamakura.

A genus which is at its best in Japan where it revels in the rich acidic, volcanic soils, often high in aluminum, and enjoys the light shade at the edges of woods, sides of streams and on the approaches to temples. A flower of the rainy season.

The young leaves of *H. macrophylla* var. *serrata* (=*H. Thunbergii*) used to be used to make ***amacha***, the sweet tea used on April 8th, the birthday of the Buddha, and as a sweetener before sugar was introduced.

H. paniculata (***nori-utsugi***) is a small tree, to 3 m, with conical heads of pinkish white flowers. The inner tissues of the bark were used as a paste in the manufacture of Japanese paper. *H. petiolaris* (***tsuru-ajisai***) is a self clinging climber with flat heads of white flowers. Both are native to Japan, along with several other species, subspecies and varieties.

INDIAN BEAD TREE

Japanese Bead Tree, Persian Lilac *(Sendan)*
Melia azedarach (Meliaceae)

A spreading, deciduous tree, up to 7 m, native of the warm coastal areas, with large bi-pinnate leaves that may reach to 1 m overall. The starry flowers are pale lilac in late spring, 2 cm across and numerous, on long stalked clusters, each flower having a prominent deep purple stamen tube. Notable for its fruits which persist through the winter, yellow, egg shaped and like bunches of beads on the bare branches.

INDIGO

Chinese Indigo *(Ai)*
Polygonum tinctorium (Polygonaceae)

An annual with smooth, reddish stems reaching 50–60 cm and green, alternate, elongated oval leaves which turn to indigo when dried. The spikes of tiny red flowers come in autumn.

It has been cultivated in Japan particularly since the 17th century. The leaves are used to produce an indigo dye for cotton cloth, the only colour which commoners were allowed to wear in premodern times.

SENDAN

AI

IRIS

Japanese Iris, Kaempfer's Iris *(Ayame, Hana-shōbu)*
Iris ensata var. *spontanea* (=*I. kaempferi spontanea*)
(Iridaceae)

Plants 60–80 cm in height bear massive flat flowers, to 20 cm in diameter, with the standards much shorter than the very broad beardless falls. Singles, doubles, shades reminiscent of orchids. The leaves, typically, have a very distinct midrib. They need a very rich soil with no lime, and are often grown in beds built up with wooden sides where the moisture can be controlled. Feeding is done during the winter, flooding only when growth is active. Flowering is the second week of June in the Tokyo area. There are hundreds of cultivars.

This Iris was introduced to Europe by von Siebold in 1858, and named after Kaempfer, who was a medical man in Japan from 1690–1692. By that time there were already masses of cultivars of the wild type, and in 1845 a samurai, Matsudaira Sakingo, had written the first basic book about them.

Iris laevigata *(Kakitsubata)*

Growing 50–70 cm in height in the wild and taller in gardens, it is a true bog plant and likes to grow with its feet in water. The leaves are light green, grassy, without a raised midrib. Typically, the flowers are a clear deep purple-blue, 10–15 cm wide and with standards nearly as long as the falls. The wild Japanese plant has given rise to many cultivars. This species will tolerate a little lime in the soil/water.

Of special interest to Iris lovers and in particular to cultivators of Kaemper's Iris and ***Iris laevigata***, are the Inner Garden of the Meiji Shrine and the East Garden of the Imperial Palace, both in Tokyo, Kōrakuen in Okayama in May/June and Kenrokuen in Kanazawa in June.

HANA-SHŌBU

KAKITSUBATA

Fringed Iris *(Shaga)*
Iris japonica

Found flowering in shady spots in the wild and in gardens from the end of April and into May. The flattish, pale-lilac flowers, 5–6 cm, with deep purple blotches and conspicuous white fringed standards, remind one of orchids. They come from a mass of grassy evergreen leaves. Each flower is short lived but there is such a succession on each stalk that the flowrering goes on quite a long time.

Roof Iris *(Ichihatsu)*
Iris tectorum

Originally from Central and S.W. China. Believed in older times to give protection from gales, it was planted on thatched roofs in country districts. Nowaday it is grown as an ornamental plant in gardens. The leaves are about 45 cm long, strongly ribbed and broad but thin in texture. The flowers, several to each stem, are a clear lilac with deeper speckles on the falls and a conspicuous white fringed crest. Flowers in May before the other irises.

SHAGA

ICHIHATSU

JUDAS TREES

Chinese Redbud *(Hanazuo)*

Cercis sinensis (=*C. chinensis*)

Judas Tree *(Seiyō-hanazuo)*

Cercis siliquastrum (Leguminosae)

The **Chinese Redbud** is a shrub 2–4 m, with bright rose pea flowers on the bare twigs in spring and even coming straight out of quite old branches. The leaves are heart shaped, 8–12 cm, bright green turning to yellow in autumn. The leaves are very similar to *Cercidiphyllum japonicum*, (page 110) but arranged alternately.

Cercis sinensis was brought to Japan from China. It grows more uprightly than *C. siliquastrum* and will stand generally more acid soils. The branches on an old *C. siliquastrum* are spreading and quite tortured in shape, and this, along with the red of the flowers, gives rise to the story that this was the tree on which Judas hanged himself. The red flowers coming out of the old wood are said to be tears of blood. Legend has it that before the crucifixion they were white.

HANAZUO

KATSURA TREE

Cercidiphyllum japonicum   (Cercidiphyllaceae)   ***(Katsura)***

A beautiful deciduous tree for all seasons which can grow to 30 m in the wild. Slightly drooping branches with oppositely arranged heart shaped leaves in iridescent shades of pink in spring, turning to green in mid-summer, then to soft, yet vivid, yellows in the autumn. Trunks of old trees are furrowed, showing a characteristic ageing. Inconspicuous red flowers on separate male and female trees.

The wood is widely used for building and implements.

KERRIA

Kerria japonica   (Rosaceae)   ***(Yamabuki)***

A bushy deciduous shrub, 1.5 to 2 m, with arching green stems, growing wild on the sides of rivers and edges of woods throughout most of Japan. The leaves are entire, toothed, sharp pointed, 6–7 cm, and a fresh bright green in spring, when the yellow flowers appear. These are usually 5 petalled, open saucer shaped, 3–5 cm wide, but are variable. Occasionally the double pom-pom of ***K. japonica florepleno*** can be found, even in the wild. This was the first form introduced to Europe in 1805 by Kerr (hence the scientific name), the single form not arriving there until 1834.

KATSURA

YAMABUKI

KONNYAKU

Amorphophallus konjac (= *A. rivieri konjac*)   ***(Konnyaku)***
(Araceae)

A strange perennial herbaceous plant which grows like a mini palm tree to almost 1 m tall, with a head of light green leaves, like those of a potato, coming from the top of a "trunk". The flowers, which appear in spring, are evil-smelling maroon aroids, speckled on the outsides, with the central maroon spadix twice as long as the spathe. It is a native of India and Sri Lanka and is cultivated in many areas, especially Fukushima and Gunma, for the fat starchy corms which provide a flour. This is kneaded with water, and heated with an alkaline solution to form a gelatinous lump. **Konnyaku** is used as a vegetable, and in **oden**. Shredded konnyaku, known as **shira-taki**, meaning white waterfall, is used in **sukiyaki**.

LACQUER TREE

Rhus verniciflua   (Anarcardiaceae)   ***(Urushi)***

A tree 6–9 m tall that is particularly noticeable in September, since it is one of the first trees to colour in autumn. The leaves, bright red and orange at this time of year, are up to 60 cm long, composed of 7–13 long pointed leaflets, glossy above, softly downy beneath. "Black lacquer" or "black Japan", as it was first known in Europe, has been used as a varnish in Japan since the Neolithic Age. Lacquer is actually made from the sap of this tree, obtained by tapping the trees every 5 days during the growing season. A tree 18 cm in diameter (5–10 years old) will produce only about 135 grams of sap in one season, after which it is normally felled. The flowers are insignificant, in drooping clusters. The yellowish berries are the source of a wax used for candlemaking.

    BEWARE! The sap can give an intense allergic reaction. Don't pick it, break off the branches, or plant it in gardens.

KONNYAKU

URUSHI

LARCH

Larix leptolepis (=*L. kaempferi, L. japonica*) **(Karamatsu)**
(Pinaceae)

A deciduous conifer to 30 m, with very straight trunk and sharp conical top. **Karamatsu** is distinguished by the round brown cones (about 2.5 cm across) which face upwards and whose scales are so much reflexed that, when ripe, they look like little wooden roses. The flowers appear in May. Leaves are in bushy tufts, 2–4 cm long, green in spring, turning golden in autumn. After leaf fall the twigs show up as reddish/straw coloured.

A fast growing and valuable tree that occurs naturally in forests in Central Honshu from 1,000 to 2,500 m and is widely planted for timber in Northern Japan. The hybrid with the European Larch (**L. decidua**) gives **L. X eurolepis**, the Dunkeld Larch, which is even faster growing.

KARAMATSU

LILIES

The Golden Rayed Lily of Japan *(Yama-yuri)*
Lilium auratum (Liliaceae)

A true aristocrat, though also a true field flower (yama=wild), growing wild in the mountains throughout Honshu in July and August. Intensely fragrant white trumpets on stems 1.0 to 1.5 m. In cultivation there may be 20–30 flowers on a stem. Six broad, white petals with a broad golden 'ray' and crimson spottings. In the wild it is said that there is one flower for each year of the plant's life, and on older plants the stem can certainly droop with the weight of the flowers. It is a stem-rooting lily with many cultivers, e.g. *platyphyllum*, with especially broad petals, and *rubro vittatum* with red rays.

The Easter Lily *(Teppō-yuri)*
L. longiflorum

A native of the Okinawa Islands, with pure white horizontal trumpets on 1 m high stems. It normally flowers in late spring, but it is a good pot plant and can easily be forced. There is a long established export trade in bulbs.

Tiger Lily *(Oni-yuri)*
L. lancifolium (=*L. tigrinum*)

A stem-rooting lily which grows to about 2 m and flowers late July. Flowers are a fiery golden orange with dark dots, of Turk's Cap form, with prominent pendulous anthers. The plants have been cultivated since ancient times, particularly for their edible bulbs. They may originally have been hybrids (possibly of *L. leichtlinii* and *L. maculatum*), and themselves are parents of the Mid-Century Hybrids.

YAMA-YURI

TEPPŌ-YURI

Showy Lily *(Kanoko-yuri)*
L. speciosum

From southern Japan. The species and its many cultivars all have flowers of pink and/or white, with combinations of bandings and spottings. The petals are broad and shallowly reflexing. In midsummer the flowers hang down most gracefully on 1 to 1.5 m stems. The fragrance is strong but not dominating. It is grown for cut flowers and is a delightful garden flower.
The bulb is orange-red to greenish yellow and edible.

Japanese Lily *(Sasa-yuri)*
L. japonicum (L. makinoi)

Grows wild in the mountains south of Central Japan. Pinkish white trumpets, usually 2–5 on a stem, held horizontally. 1 m high, in May to August. Sweetly scented.

L. rubellum *(Himesa-yuri, Otome-yuri)*

Very similar to *L. japonicum* but smaller, with shorter fatter leaves and fewer flowers. The flowers are pink, open trumpets, held horizontally, 0.5 m high. May. (**hime**=small)

L. maculatum *(Sukashi-yuri)*

Grows wild, especially on cliffs and beaches in Northern Japan. Open, upright, pale orange trumpets on stems about 50 cm high from May-August. Many varieties and forms. A parent of the *L. xmaculatum* hybrids which are the most tolerant of all garden hybrids.

Oriental Hybrids owe much to Japanese species, in particular *L. auratum* and *L. speciosum*, and to the Chinese *L. henryi*.

KANOKO-YURI

SASA-YURI

HIMESA-YURI

SUKASHI-YURI

LOQUAT

Eriobotrya japonica   (Rosaceae)   **(Biwa)**

A round evergreen tree, native to China and Japan, of 6–9 m height with leaves suggesting wrinkled rhododendron leaves with brown felting on their undersides. The yellowish-white flowers are only about 1 cm across but are carried on stiff pyramidal clusters with brown hairy stems. The fruits come in early summer, woolly, yellow, the size of a plum but with the taste of an apple x pear, with 3–4 brown seeds at the centre. They are a popular fruit in Japan, and have been the subject of breeding and improving. As the flowers and fruitlets are carried through the winter, it is understandable that the most favoured areas for its cultivation are in the mild coastal areas of the south, but it can be found in most parts of the country where the winters are not too severe. The fruits may be eaten raw, stewed or preserved.

The Japanese name is said to be taken from the three stringed instrument, the **biwa**, which the fruit resembles in shape. It is usually planted in temples and shrines rather than private gardens, since the spreading branches and dense evergreen foliage create a damp and dark area where few plants flourish.

BIWA

LOTUS

Nelumbo nucifera (Nymphaceae) *(Hasu)*

A true water plant with ancient and famous associations. Simple round waxy leaves on stalks 30–90 cm, stand out of the water like umbrellas. The flowers are in shades of pink, like long stemmed water lilies, but with prominent ovaries in the centres looking like the rose of a watering can, which when ripe hold brown seeds. Flowering is in the hot, humid days of August.

The roots, rhizomes of 8 cm diameter and over 1 m long, are a staple vegetable known as **renkon**.

The Lotus has close associations with Buddha and the Buddhist faith. There are often Lotus beds near temples and frequent carvings and paintings of both flowers and leaves within temples. To sit on the Throne of Hasu (The Lotus Throne) means rebirth in paradise after death, and is a goal of the Buddhist faith. Tombstones are often decorated with carved lotus leaves and flowers and carvings, sculptures and paintings are often used in temples.

Many years ago, before so much land was under concrete and agricultural land became better drained, lotus was a much more common plant and children would enjoy eating the seeds.

See them at the Shinobazu Pond in Ueno Park in Tokyo and in still water throughout the country. The flowers open each day at dawn and close by noon, so early morning visits are required to see them at their best. Each flower lasts only 2 or 3 days but the beds are in flower for about 6 weeks through the hottest part of the year.

HASU

RENKON

MAGNOLIAS

The Northern Japanese Magnolia *(Kobushi)*
Magnolia kobus (Magnoliaceae)

A harbinger of spring on the leafless hillsides in March/April, where the white flowers, on deciduous trees to 10 m, can be mistaken at a distance for piles of snow. All magnolias have simple, untoothed leaves, alternately arranged. *M. kobus* has light green pointed ones to 10 cm long, white on the undersides. The flower buds are distinctively pointed, greyish and hairy, and the shape of them has given rise to the Japanese name—**kobushi**=fist. The flowers are about 10 cm across, made up of 6 white strap-shaped petals and 3 petalloid sepals, which together look like 9 petals. The fruits are pink, 7–10 cm long, splitting all over to release red seeds which dangle by threads before dropping. They are used in **ikebana**.

The Star Magnolia *(Shide-kobushi)*
M. stellata

Often regarded as a garden variety but in fact grows wild in Central Japan making a shrub of 2–4 m. It is similar to **M. kabus** but the fragrant flowers have 12–18 'petals', narrower than **kobushi**, and with many of them curling back when they are in full flower in early spring. Leaves appear later.

There are many popular and beautiful cultivars and hybrids of these two species, as there are of many of the Magnolias, especially in temperate areas thoughout the world with soils with high organic matter, adequate moisture, and where frost will not damage the flower buds.

KOBUSHI

SHIDE-KOBUSHI

The Lily Flowered Magnolia   *(Mokuren)*

M. liliflora

A deciduous shrub 3-5 m, with straggling branches. The flowers are large, upright chalices, darker on the outsides, fading to pink on the insides, flowering as the leaves develop in mid-spring. A parent of *M. x soulangeana*, and *M. x Lennei*.

Yulan   *(Haku-mokuren)*

M. denudata

Introduced from China. Tree 4–6 m bearing white flowers in March/April. Sometimes this is incorrectly called the Tulip Tree.

Japanese Cucumber Tree, The Big Leaved Magnolia   *(Hō-no-ki)*

M. obovata   (=*M. hypoleuca*)

A big, deciduous tree to 25 m, found in mountain areas throughout Japan. Big, leathery leaves, 20–40 cm long, cluster at the ends of the branches and surround scented, creamy white flowers which resemble water lilies. These striking flowers, 15 cm across with a dark brown central boss and prominent stamens, appear in May-June.

The fine grained wood, being easy to carve, was used in blocks for wood block printing and for **geta** and sheaths for samurai swords. The leaves of most magnolias skeletonise well and are often seen in flower arranging.

If you visit Hida Takayama in the mountains of Central Japan, you will find this leaf used as a pan for roasting **miso**.

MOKUREN

HŌ-NO-KI

MAPLES

(Aceraceae) *(Momiji)*

There are many species native to Japan. Most cultivated varieties are hybrids and cultivars of *Acer palmatum* and *Acer japonicum*. Most have insignificant flowers. In 1710 there were 36 named varieties listed. By 1882 there were already more than 202.

The Downy Japanese Maple or Full Moon Maple
Acer japonicum *(Hauchiwa-kaede)*

Has downy leaf stalks and slightly downy young keys. The leaves are often quite rounded and the lobes are not very deeply cut. The wings of the seeds are widely spreading. It occurs wild on mountain slopes at about 1,000 m. It is frequently used in bonsai as it is quite a tough subject and is colourful both in spring and in autumn.

The Smooth Japanese Maple *(Iroha-kaede)*
A. palmatum

The leaves are 5–7 lobed and toothed. The lobes are cut more deeply than in **A. japonicum**. It is a tall tree, to 10 m, found wild in mountain areas, except it Hokkaido.

The cultivated palmatum hybrids are often divided into 3 groups: those with 5-lobed leaves, those with large, generally 7-lobed leaves, and those with dissected and even thread-like leaves of 7-11 lobes.

KOHAUCHIWA-KAEDE

IROHA-KAEDE, A CULTIVAR

Also native to Japan are *A. carpinifolium* with leaves easily confused with Carpinus (Hornbeam) or Zelkova, but making a small tree with spreading branches, *A. crataegifolium* with unusually thick leaves, in shape like a hawthorn, *A. miyabei*, a small tree with palmate leaves, yellowing in autumn but keeping a red leafstalk. *A. nikoense*, the **Nikko maple** (*Megusuri-no-ki*) and *A. cissifolium* (the **Vine Leaved Maple**) both have leaves with three leaflets, more deeply toothed in *A. cissifolium* which has keys hanging in long bunches. The keys of *A. nikoense* are in threes and are downy. *Acer rufinerve* has the descriptive English name of **Grey-budded Snake Bark Maple**. The **Red Snake Bark Maple**, (*A. capillipes*) with its green and white snake bark, usually has red leaf stalks and always has brilliant red leaves and fruits in the autumn. The shape of the leaves of *Acer buergerianum* is explicit in its English name—the **Trident Maple**. It is a favourite bonsai subject.

Leaves of maples are always arranged oppositely on the stem—this fact helps prevent confusion with such genera as *Liquidamber*.

Maples are the main ingredients of the Japanese autumn, whether brocading the mountain sides, or as individual trees strategically placed in gardens to catch the slanting sunlight, or even as tiny bonsai. In mid-October, Nikkō, and its surrounding hillsides, will be ablaze. In early November, viewing will be at its height in Kyoto. The colour will be everywhere, but Arashiyama is a 'must'. The glen at the Tofuku-ji Temple is breathtaking, and at Katsura Villa the maples rival the azaleas of spring time for their fire.

MARE'S TAIL (Horse's Tail)

Equisetum arvense (Equisetaceae) *(Tsukushi)*

A familiar plant with edible shoots that appears on uncultivated land and at the edges of fields in March and April. It consists of fertile, pinkish-brown stems bearing collars of scales and a cone-like head of spores at the top. About 10 cm tall. Later sterile stems will appear with whorls of green, resembling leaves, at each joint giving the plants the appearance of mini-Christmas trees. Hollow stems reaching 30 cm.

A very spreading and invasive plant. The young shoots can be eaten simmered in soy sauce and sugar.

TSUKUSHI

MORNING GLORY

Pharbitis nil (=*Ipomoea nil*)   (Convolvulaceae)   **(Asagao)**

Bright blue, violet, or rose trumpets in summer on an annual vine, climbing by its twisting stems. The green leaves are pointed hearts, or gently lobed arrows, soft in texture. Growing wild, or in gardens, or, since the 18th century as a 'cult' culture in pots. This pot culture has had a fluctuating popularity; there have been shows, competitions and much hybridising, with seeds selling for vast sums. It reached its zenith in the early 1900's. Early morning visits were popular, although you had to be up early, for the flowers had faded before noon. "Best seen at 4 am of a scorching July or August day—the flowers unfolding make you forget the heat of the day that is to come"—said Florence du Cane in 1908.

There is also the "Daytime Flower", or **Hirugao** (*Calystegia japonica*) which is very similar, but, which as its name would suggest, keeps different hours. It is usually pink.

You can see and buy potted Morning Glory at the Asagao Ichi (market) held every year at the Kishibojin Temple, Tokyo from 6th to 8th July.

ASAGAO

MULBERRY

Morus bombycis (Moraceae) *(Kuwa)*

A deciduous tree that will grow up to 10 m in the wild throughout most of Japan, with heart shaped, or 3-lobed, leaves. However it is more usually seen growing in bush form for use as food for silkworms. The leafy shoots are cut off each year as soon as the leaves mature and then repeatedly until the fall. The shoots, which are about 1.5 m long, are fed to the silkworm larva. This cropping gives a coppice effect to the bushes and a stool at a height of anything from 0.5 to 1 m from the surface. The plantations are usually grown from cuttings and are often intercropped, especially whilst the bushes are still young. Research on the cultivation and hybridising of mulberries and on the breeding and rearing of silkworms has made a tremendous improvement in the economy of silkworm production since the Second World War. Silk of very fine quality is produced in Japan but the proportion of the market supplied by home grown silk is now quite small compared to former days.

The black juicy berries arouse nostalgia in those Japanese who spent their childhood in the country.

KUWA

OSMANTHUS

Sweet Olive *(Gin-mokusei)*

Osmanthus asiaticus (=*O. fragrans, O. latifolius*)
 (Oleaceae)

Originally from China, now a favourite garden shrub 3–4 m, with pleasing dark evergreen leaves to 10 cm. The small white flowers appear in short stalked clusters in the leaf axils in late autumn. The insignificance of the flowers belies their intense sweet fragrance, which is the real reason for growing this shrub. A fragrance which has been described as a blend of jasmine, gardenia and ripe apricots, and which comes as a pleasant surprise as the days grow colder and shorter.

Osmanthus fragrans var. *aurantiacus* *(Kin-mokusei)*
 (=*O. aurantiacus*)

Very similar to the above but with deep golden flowers. A very popular plant which is hardy north of the Kanto.

Osmanthus fragrans var. *thunbergii* *(Usugi-mokusei)*
 (=*O. aurantiacus* var. *thunbergii*)

Has flowers tinged with yellow or light orange and is often confused with **Kin-mokusei**, but it is not so hardy and is planted in the warmer regions of S.W. Japan.

O. ilicifolius (=*O. heterophyllus, O. aquifolius,* etc.) is native to Japan, has leaves to 3 cm with several spiny points similar to holly leaves. The young leaves are not toothed. It has fragrant white flowers in autumn. ***O. fortunei*** with broader spiny leaves to 6 cm is often considered to be a hybrid of *O. ilicifolius*. *O. delavayii* is from China, flowers in spring, has small spiny leaves and will enjoy lime in the soil.

In warm climates **Osmanthus** shows its relationship to Olive by forming dark purple black fruits, but most **mokusei** plants in Japan are male.

USUGI-MOKUSEI

KIN-MOKUSEI

PAEONIES

The Tree Paeony, The Moutan *(Botan)*
(Paeoniaceae)

Paeonia suffruticosa. Also cultivars based primarily on the hybridising of *P. suffruticosa* (very fragrant rose pink to white flowers with maroon central blotches), *P. lutea* (yellow, single, up to 1 m), and *P. delavayi* (up to 2 m high with single, dull red, flowers).

A very ancient plant in China, it was brought to the temples of Japan in the second half of the 8th century at the same time as Buddhism. It was used chiefly for medicinal purposes (the root is still used in Chinese medicines), but also grown in the courts for the beauty of its flowers. Consequently it is the subject of the most exquisite paintings on scrolls and on screens. In the 10th century 39 varieties were on sale in China and there was a big trade in the wild species for rootstocks for grafting—still the usual way of propagation. These hybrids were already fully double when they came to the R.H.S. Gardens in England in the mid-19th century, via the great collector, Robert Fortune.

To produce the massive, heavy, fragrant flowers (at least 30 cm diameter) they need deep rich soil, generous mulching, and plenty of moisture in the growing season after flowering. They like cold winters which give complete dormancy, but must not have late spring frosts on their delicate new foliage. Cultivars are propagated by grafting, using rootstocks of **Shakuyaku**.

Herbaceous Paeonies *(Shakuyaku)*

Grown in variety, especially the Anemone Flowered one, but not so highly thought of as the **Botan**.

P. japonica (single white), and *P. obovata* (single purple) grow wild in the woods of the cooler areas. Both are herbaceous, 30–40 cm high.

BOTAN

SHAKUYAKU Old Japanese saying: A beautiful woman should stand like a paeony, sit like a tree paeony, and walk like a lily.

PAGODA TREE

Sophora japonica   (Leguminosae)   ***(Enju)***

Native to China, where it is known as the "Tree of Success in Life". Often grown as a street tree, but if not clipped can grow to 20 m. Leaves to 30 cm long are compoundly pinnate with 9–15 leaflets and whitish undersides. Stiff erect clusters of creamy white, pea-shaped flowers appear in mid summer and give rise to hanging pods of fruit which ripen to yellow in September/October. These are conspicuously constricted between the seeds, suggesting a rosary, and the seeds themselves are like sticky beads.

ENJU

PALMS

Chinese Windmill Palm, Chusan Palm *(Tōjuro)*
Trachycarpus fortunei (=*T. excelsus, Chamaerops fortunei* and many other synonyms) (Palmae)

Originally from Central and Southern China and often found in gardens, especially in Kansai (Kyoto-Osaka area). A single, shaggy, fibrous trunk can grow to 9 m. The flat fan-like leaves clustered at the top of the trunk are up to 1 m wide, and divided into numerous stiff segments on long toothed leaf stalks. The small yellow flowers hang down in clusters amongst the leaf bases—male and female on separate trees. Fruits are yellow at first, and later become blue/black.

Trachycarpus excelsa (and many synonyms) ***(Shuro)***

This was originally a native of Southern Kyushu but is now only found in gardens, mainly in the warmer areas. It is distinguished from the ***Tōjuro*** by having the long leaflets bent over, much as the fingers on a hand, or the prongs of a bamboo rake, bend over.

SHURO

DWARF PALMS
Large Lady Palm, Dwarf Ground Rattan *(Kannon-chiku)*
Rhapis excelsa (=*R. flabelliformis*)

Slender green canes, usually below 1 m when grown as an ornamental plant, bearing dark, glossy, leathery leaves to 60 cm, each with 5–10 chubby 'fingers' and looking like a hand or a **sasa** leaf. (**chiku**=bamboo)

Rhapis humilis *(Shuro-chiku)*

Similar to the Lady Palm, but taller, 4–5 m, and more graceful. The leaflets are more numerous, longer and more slender. The trunks retain a coarse netting. It is often seen in tubs.

Together, these two palms are called **Kanso-chiku**. Around 100 varieties have been developed which are defined according to the variations in the stripings on the leaves. They are some of the plants grown in pots as **Koten engei**, the products of a unique and somewhat esoteric method of growing highly modified varieties of traditional plants, especially those whose leaves show variations e.g. *Rohdea japonica* (**Omoto**) and *Asarum spp.* (**Wild ginger** or **Saishin**). The varieties have evolved and multiplied since the 17th century and are yet another facet of the Japanese attention to detail. They are usually grown in glazed pots that have three curved feet.

KANNON-CHIKU

SHURO-CHIKU

PAULOWNIA

Paulownia tomentosa (Scrophulariaceae) **(Kiri)**

A quick growing deciduous tree, reaching 12 m in 20 years, and often coppiced. It grows wild in the mountains of southern Honshu and Kyushu, but opinion is divided as to whether it originated in China or Japan. The leaves are opposite, triangular to heart shaped, the larger ones being 3–5 lobed and up to 30 cm long. The flowers come in April/May, before the leaves, and are like mauve foxgloves in upright conical clusters. The fruits are 3–4 cm at first pointed and green, but splitting to release seeds with circular wings.

The wood is light and fine grained and a **Kiri** tree was traditionally planted at the birth of a baby girl, so that it might be cut down to make a chest of drawers for her dowry. **Kiri** wood is suited to this purpose as it does not warp with charges in humidity and acts as a moth killer. It is good for making high quality **geta** and special boxes. It has a long history and was mentioned in the Tale of Genji.

KIRI

PEACH

Prunus persica   (Rosaceae)   ***(Momo)***

In spite of its specific name it is a native of China where it commands the respect the Cherry enjoys in Japan. It has been cultivated since ancient times; peach seeds were found in pots of the Yayoi period (200 BC–200 AD), and records from the early Tokugawa period (1600s) identity many varieties and their areas of cultivation. Nowadays they are grown extensively in the Okayama district where humidity and rainfall are not too excessive and, for later crops, as far north as Fukushima. Varieties suitable to the climate have been developed.

Flowering Peaches   *(Hana-momo)*

Many varieties developed as ornamentals, including the Weeping Peach (**Shidare-momo**) and the Chrysanthemum Flowered Peach (**Kiku-momo**). One sometimes sees grafted trees with both pink and white flowers—red and white being a combination that suggests happiness and joy in Japan.

The story of Momotarō, the most famous of Japanese nursery stories, tell of a little boy, born from a large peach stone. Momotarō, the peach boy, grew to be strong and brave, and having killed the devils on Devil Island, brought back great riches to his adopted parents. The Peach blossom is an integral part of the Girl's Festival on 3rd March (**Hina-matsuri**, the Festival of Dolls) and is used in ***ikebana*** at this time, often with rape flowers.

MOMO

HANA-MOMO

PEARS

The Japanese Pear *(Nashi)*
Pyrus serotina, and its cultivars and hybrids   (Rosaceae)

A native pear that has been very greatly developed to give many cultivars and hybrids. The leaves are typical pear leaves, long pointed ovals to 7 cm, and the flowers are typical of pears—white open saucers to 4 cm in clusters along the branches. BUT ... the fruits look like a beautiful russet apple. They are flattened and perfectly round, and well grown cultivars will be at least 10 cm across. They are in the shops from early September, and many varieties will keep for 8–10 months from harvest.

The trees can be grown even in the warmer parts of Japan as they need little winter chilling. They are usually grown quite closely planted in orchards with the branches trained on a horizontal overhead trellis 1.5 to 2 m high. High yields are thus obtained and the trellis supports both branches and fruit, protects against the wind, and makes all operations on the fruit easier. Fruits are thinned very carefully in the early stages, then enclosed in bags of waxed paper, or even of newspaper, to give protection from the weather and the many insects. The resulting fruits are large and perfect, designed to be peeled and sliced, not bitten!

Pears (Western Pears) *(Seiyō-nashi)*
Pyrus communis and its cultivars

This is more suitable for cultivation in the cooler parts of Central and Northern Honshu. It has a short season in late autumn, and is not so widely used as **nashi**. The fruit of some cultivars tastes better eaten in early winter.

NASHI

PERSIMMONS

Japanese Persimmon *(Kaki)*

Diospyros kaki (Ebenaceae)

Native, small rounded deciduous tree to 10 cm, conspicuous in late autumn when the fruits, resembling oranges in size and colour and tomatoes in shape, hang on the bare dark brown branches. In summer the entire, pointed, dark green leaves are to 8 cm. The flowers, insignificant and greenish, appear in July.

The type known as **ama-gaki** are sweet and can be eaten raw. They are at their best when almost over ripe. **Kaki** thrive in a climate similar to that needed by the Satsuma orange, and like the Satsuma, they are grown over a large area of Southern Honshū, though they can also be grown further north. Though they are grown in many countries, the best fruits are reputedly produced in Japan. They are certainly widely grown and greatly enjoyed. The type known as **shibu-gaki** are astringent and cannot be eaten raw from the tree. They will grow in wider climatic conditions, but especially where autumns are dry and warm. They are dried, peeled, and hung in strings under the eaves of houses in the country districts to dry. They then become like dried figs and are quite delicious. Alternatively they can be rubbed with alcohol, wrapped individually in papers and stored for ten days in a box to lose their astringency.

The infused leaves make a healthy drink, rich in vitamin C, and the young leaves are sometimes used to wrap **sushi**, and as a garnish.

KAKI

PHOTINIA

Japanese Photinia *(Kaname-mochi, Akame-mochi)*
Photinia glabra (Rosaceae)

Native to southern Japan and East China, makes an evergreen tree to 2.5 m, with smooth, leathery, oval pointed leaves 5–8 cm long. The small whitish-pink flowers are in clusters 5–10 cm across followed by hawthorn-like berries.

P. glabra rubens has young foliage of bright red, and pink flowers.

Chinese Photinia *(Ōkaname-mochi)*
P. serrulata

Not native to Japan, though found in gardens. Can grow to a small tree with smooth bark. The leaves are toothed, especially on the younger shoots, with grooved leafstalks. Flowers white, in flattened clusters like *P. glabra*. Some reddish new leaves in spring, and again in autumn.

Photinia x fraseri is a hybrid of these two which arose as a chance seedling and has particularly vivid bronze-red new growth in spring. It is most often seen as a hedge, the clipping producing constant new red growth, but reducing the flowering and fruiting. Many cultivars.

AKAME-MOCHI

PIERIS

Andromeda (the old generic name)

These evergreen shrubs/small trees have tapering shiny leaves to 7 cm, all along the branches, but they are concentrated in whorls at the ends of the branches. The young leaves are often bright red in spring, when the white or pink flowers hang in clusters like lily of the valley. There are many cultivars.

Chinese Pieris
Pieris formosa   (Ericaceae)

Grows to 5 m and is from S.E. China and N.E. Burma. It is cultivated in gardens, but is understandably not so hardy as the Japanese Pieris, below.

The Japanese Pieris *(Asebi)*
P. japonica

A narrow bush to 3 m with leaves which tend to be thinner both in width and texture. The flowers, March to May are also in longer, thinner panicles than in *P. formosa*. It grows wild in the forests of the warmer parts. There are many cultivars including the variegated *P. japonica variegata* with yellowish white edges to the leaves and pinkish young growth.

The leaves are poisonous, and are not eaten by deer, which explains how they survive in the parks at Nara.

ASEBI

PINES *(Matsu)*

The Japanese Red Pine *(Aka-matsu)*
Pinus densiflora (Pinaceae)

A tree, to 35 m distinguished by its red trunks with irregular plates of bark, needles in groups of 2, male flowers in dense clusters at the base of the new wood in April and oval cones to 6 cm. An upland forest tree used for general construction work, and for pulp. Long cultivated in gardens and for bonsai.

The Japanese White Pine *(Goyō-matsu)*
Pinus pentaphylla (=*P. parviflora*)

A slow growing tree in gardens, but to 25 m in the wild, with grey bark. Slender needles are slightly twisted and are in groups of 5, cones are oval-pointed, to 6 cm, in persistent clusters. Its easily trained wood makes it a favourite bonsai subject.

The Japanese Black Pine *(Kuro-matsu)*
Pinus thunbergii

A strong tree in every sense. To 35 m in the wild, with dark grey, fissured bark, dark green, stiff needles in 2's, broad cones to 6 cm with scales with little spines. A strong rugged shape whether in the wild, in gardens or in bonsai.

Parasol Pine, The Japanese Umbrella Pine *(Kōya-maki)*
Sciadopitys verticillata (Taxodiaceae)

A medium sized regular cone-shaped tree in cultivation, but growing to 15 m in its native S.W. Japan, with reddish bark peeling off in strips. Slender needles (8–12 cm) grow in whorls like the spokes of an umbrella. The roundish cones (5–7 cm) easily break up on handling.

KA-MATSU

GOYŌ-MATSU

RO-MATSU

KŌYA-MAKI

PINES: The Background

Pines are a quintessential part of Japan; of the landscape, the legends, the paintings and the traditions. Large ones form a serene background to temples and tea houses, small ones on island cliffs seem to be the inspiration for the bonsai on city balconies. Ever green, the pine symbolizes long life and constancy. It is indispensable to the New Year ikebana and to **kado-matsu**, stylized decorations of Pine, Bamboo, and Plum placed either side of the gate, which perhaps give rise to the Japanese name for the New Year season "Matsu no uchi" or "within the pines". A painting of a gnarled pine tree forms the permanent backcloth of the Noh stage.

The "Pine-Clad Islands" of Matsushima had so much impact that even the poet Bassho said he was at a loss for words to describe their beauty. Even in these more prosaic days one often trains a pine tree to hang over the garden gate.

There are beautifully trained pine trees in parks and gardens all over Japan, some of the most striking in their simplicity being the rows of them standing firm on the grass on the south side of the Imperial Palace, Tokyo. An appropriate place for the king of trees.

PLANES

Platanus spp. (Platanaceae)

Deciduous trees, often found in streets and parks, particularly noticeable by the flaking patches of bark giving a creamy marbled effect. The hanging clusters of spiky fruits on the bare branches in winter only need frosting to make them look like Christmas tree decorations. The leaves are palmate and lobed, about 20 cm across, turning golden yellow in the autumn. Excellent city trees (apart from the considerable debris from autumn leaves and late winter fruits) because they withstand pollution, pollarding and cramped roots. Male and female flowers are insignificant, in separate clusters on one tree.

The London Plane   *(Momijiba Suzukake-no-ki)*
Platanus x acerifolia (=*P. hybrida*)

Probably a hybrid of *P. orientalis* and *P. occidentalis*. The most widely planted. Grows to 9 m in 25 years, eventually to 30 m. The leaves are deeply 3–5 lobed, though very variable. Flowers and fruits hang in pairs.

American Plane, Buttonwood, Buttonball
P. occidentalis   *(Amerika Suzukake-no-ki)*

Slightly more vigorous than the London Plane. Leaves shallowly lobed.

The Oriental Plane   *(Suzukake-no-ki)*
P. orientalis

Native to Eastern Europe. Eventually will grow to 25 m or more, making a rounded, spreading tree. The leaves are much more deeply cut and divided into 5 or 7 lobes. Fruits hang in bunches of 3–6.

SUZUKAKE-NO-KI

PODOCARPUS *(Maki)*

The Yew Podocarp, The Japanese Yew   *(Inu-maki)*
Podocarpus macrophyllus   (Podocarpaceae)

A narrow, upright, evergreen shrub or small tree, eventually to 20 m. The glossy dark green leaves are similar to pine leaves, but are flatter and longer, 7–15 cm long, with a distinct midrib. The male flowers are in clusters of stumpy yellow spikes. The female are green and inconspicuous, on separate trees. The round green seeds with a fleshy covering are 2 cm in diameter and are ripe in October. Over the winter they turn black, become more egg shaped, and can be eaten.

A tree that lends itself to clipping and training in gardens since the spirally arranged leaves then appear to stand upright.

INUMAKI

QUINCE

(Often called 'Japonica' for its common, or English name, and misnamed 'Cydonia' for its Latin name)

The Japanese Quince *(Boke)*

Chaenomeles lagenaria (Rosaceae)

A deciduous spreading shrub to 2.5 m which originally came from China. Leaves are oval, toothed, glossy, to 8 cm, on spiny stems. The flowers are open cups of 5 rounded petals clustered along the stem in spring and may be any colour from dark red, through pink and white to bicolored. There are around 30 garden varieties and it is popular for spring ikebana. The fruit is about 8 cm long, an elongated oval.

Maule's Quince, Lesser Flowering Quince *(Kusa-boke)*

Chaenomeles japonica

Grows wild in Japan, in mountains and woods, and is often seen on banks flowering amongst the grass in April/May. It makes a low spreading evergreen shrub to 50 cm in the wild, but grows taller in gardens. The twigs are spiny and warted, and bear spoon shaped leaves to 5 cm. The flowers are brilliant orange cups, 2.5 cm across, followed by hard yellowish apple-like fruits about 3 cm across. Grown as an ornamental.

BOKE

KUSA-BOKE

REEDS OR RUSHES

Mat Rush, The Tatami Reed *(I)*

Juncus effusus var. *decipiens*   (Juncaceae)

Like most reeds, this is a perennial that enjoys damp places. It is cultivated in wet fields, particularly in Okayama (Southern Honshu). It grows from 0.5–1.5 m, with round smooth green stems and reddish, undeveloped leaves at the base. The brown flowers, in summer, are in dry clusters, one cluster fixed tightly to each stem. The reeds are harvested in spring, dried, then woven to make the silky-smooth surface of the **tatami** mat. (The inner part of the mat is made of compacted rice straw.)

1

RICE

*(**Ine**=Rice plant)*
*(**Kome**=Rice grain)*
*(**Gohan**=Cooked rice)*

Oryza sativa (Graminae)

A cereal grass. Its inflorescence is a loose panicle. Its seeds being the hard grains of rice. Seedlings grow rapidly in nursery beds under plastic tunnels, and after about 30 days will be ready for planting out, formerly by hand, increasingly by machines. The fields are rotavated in a boggy condition, then shallowly flooded and the planting done under water. In the colder parts of Northern Honshu only one crop per year is possible, but further south it forms just one crop in a seasonal cropping system, with other grains or with vegetables. In the very south, two crops per year are possible. In the paddy fields, the water level is lowered after flowering until it is dry for harvest.

'Dry' or 'upland' rice grows on moisture retentive soils in areas where there is adequate summer rain to make flooding unnecessary. Different areas have their own individual ways of getting the harvest home and dry. Sometimes on racks, sometimes stooked—the architecture of the stook varying from district to district. Increasingly it is being combined, the small wet fields demanding special small, tough combine harvesters.

Although fewer people now eat three meals of rice a day, rice is still intrinsic to the culture. The word for a 'meal' is synonymous with the word for cooked rice-**gohan**. **Mochi** rice is a glutinous variety of rice which is steamed, pounded and made into cakes essential to the New Year cuisine. Drink, too, is made from rice. **Sake** drunk warm, is a very easily acquired taste. It is produced from the best of the rice crop, fermented like wine, giving a final strength of over 15°.

INE

SAFFLOWER

Carthamus tinctorius **(Benibana)**

An annual to 1 m with alternate, hard, dark green, prickly leaves and bright orange thistle-like flowers. These are 2.5–4 cm in diameter and later turn red.

The young flowers are chemically treated to wash out the yellow pigment. The red pigment which remains formed an important pink dye for silk, and for make-up (hence **kuchibeni**= lipstick). A crop which is not grown as much as it was, but which is enjoying a revival, especially in Yamagata Prefecture.

SANSHO

Japanese Pepper, Japanese Prickly Ash **(Sanshō)**

Zanthoxylum piperitum (=*Xanthoxylum piperitum*) (Rutaceae)

A deciduous mountain shrub 3–5 m, with pinnate leaves, 5–15 cm, made up of 11–19 leaflets and with two short sharp thorns at the leaf bases. The flowers, in April/May, are in small, yellow, loose clusters on separate male and female plants. The fruit has a red covering in October, and splits to reveal a black pip which, when ground, makes a spice which can be sprinkled on grilled eel or grilled aubergines. A young leaf is often used as a garnish for spring food, and the leaves chopped into **miso** make 'green miso' which is served with charcoal grilled tōfu (**dengaku**).

The wood is used to make the traditional mortar that is used for grinding spices and herbs.

BENIBANA

SANSHŌ

SENRYO

Chloranthus glaber *(Senryō)*

A low-growing evergreen shrub of 50–80 cm, found in forests in the warmer parts. The leaves are thin and bright green, yet quite leathery and smooth. They are 6–14 cm long, pointed ovals, with deep toothing at the ends. Two or three, stalked, yellowish-green clusters of flowers appear in summer at the ends of the branches give rise to red berries (rarely yellow), which ripen in winter and are favourites for New Year flower arrangements.

"**Senryō** = 1,000 **ryō** (an old gold coin), and along with **Manryō** (=10,000 **ryō**, see page 16) is considered to bring money and riches to households in the New Year.

SENRYŌ

SHISO

Perilla, Shiso *(Shiso)*
Perilla frutescens (Labiatae)

Native to South and Central China. An annual of 20–40 cm. It is a typical labiate plant—a family that gives so many fragrant culinary herbs. Branched square stems bear opposite leaves on long stalks. Leaves are broadly ovate with pointed ends, 6–8 cm long, much toothed, thin, soft and fragrant. The flowers are in long narrow spikes in summer and autumn. The red variety, **Aka-jiso**, has reddish-purple leaves and purple flowers and in Japan is used as a rich but natural crimson coloring for **umeboshi**—pickled plums. In the West it is gaining popularity as an ornamental. The green variety, **Ao-jiso**, has highly fragrant leaves and white flowers. The leaves are used widely as a herb: fried in **tempura**, rolled in **sushi**, and chopped over **tofu** and salads. The little seeds in their casings are salted and used in making pickles to give the **shiso** flavour in winter.

AKA-JISO

AO-JISO

SILK TREE

Persian Acacia, Pink Siris Tree *(Nemu-no-ki)*
Albizzia julibrissin (Leguminosae)

Found wild in forests in southern parts of the warm temperate zone, and as a street tree and ornamental in warmer parts. A graceful spreading deciduous tree up to 9 m. In the wild the trunk usually slopes, and the branches grow horizontally from it. The ferny bi-pinnate leaves droop and close up at night, giving it its Japanese name of "Sleepy Tree". The pink feathery pom-poms that are really the flowers, come in the rainy season. They are actually bunches of stamens with small petals at their bases. The fruits are flat pods, up to 15 cm long, that hang from the branches long into the winter.

NEMU-NO-KI

SNOWBELL TREE

Styrax japonica (Styracaceae) **(Ego-no-ki)**

A native deciduous tree, 3–5 m, growing all over Japan, with graceful spreading branches. Noticeable in early summer for the clusters of white flowers on short lateral shoots massed along the undersides of the branches. Five petals make a bell, with a clapper of prominent yellow stamens. The fruits are a characteristic long-stalked green berry, slightly egg shaped, with the calyx making a snug fitting hat. The leaves are entire, dark glossy green, to 8 cm.

Big Leaf Storax **(Hakuunboku)**

Styrax obassia

Also native, and similar, but making a taller, narrower tree, to 9 m. The 2–3 cm white flowers hang in slender racemes, 15–20 cm long. The leaves are more round than in ***S. japonica***, and are at least 10 cm long and are downy underneath. The fruits have a velvety down on them.

The wood of Styrax is very hard. It is used in umbrella making to make the end into which all the bamboo fits. It is the traditional material and stands the wear of being constantly put up and down.

SPIDER LILY

Equinox Flower *(Higan-bana)*

Lycoris radiata (Amaryllidaceae)

Scarlet trumpets, spidery because of the long stamens and pistils and narrow curved perianth segments, cluster at the top of fleshy stems (scapes) up to 30 cm tall. There are no leaves at this time of the year—around the autumn equinox—they will come later, and be strap shaped, dark green, coming straight from the bulb, which is poisonous.

Higan=Autumn Equinox, which, like the spring equinox, is an important Buddhist festival.

Found at the edges of fields, on banks, and around the moat of the Imperial Palace in Tokyo. It is an ancient belief that **Higan-bana** is one of the flowers found in Paradise, and so it is often planted in graveyards.

SPIRAEA

(Rosaceae)

A genus with many native species widespread throughout Japan which have given rise to many cultivars throughout the world. The following are some of the more well known species.

Spiraea thunbergii *(Yuki-yanagi)*

Originally a native deciduous shrub found in gardens or forced for cut flowers in winter and early spring. The arching branches of 1–2 m have small, narrow leaves and in early spring are covered with massed clusters of small white flowers. A parent of the Bridal Wreath (*S. x arguta*).

S. japonica *(Shimotsuke)*

A low growing deciduous shrub to 1 m, found wild in woods and open spaces throughout Japan. The leaves are narrowly oval and pointed, toothed and whitish on the backs. The flowers appear in June in flat pink heads and are slightly fragrant. It has given rise to the cultivars *S. bullata*, and *S. Anthony Waterer*.

YUKI-YANAGI

SHIMOTSUKE

S. nipponica **(Iwa-shimotsuke)**

Has a dense bushy habit, to 2 m, small oval leaves about 1.5 cm, and clusters of flat heads of white flowers along the uppersides of the branches in May/June. ***S. nipponica*** var. ***rotundifolia*** has slightly larger rounder leaves, and the flowers come in flat heads at the ends of the branches and side branches in May. They are white, and slightly larger than the type plant.

S. salicifolia **(Hozaki-shimotsuke)**

Comes originally from Northern Japan, where it grows to 2 m, has willowlike leaves and coarse terminal spikes of pink flowers in summer.

Reeves Spirea **(Kodemari)**
S. Cantoniensis

Grown in gardens and for cut flowers, **Kodemari** has been a popular shrub since the Edo period. It is deciduous and its graceful, spreading stems reach 1–2 m. Rounded clusters of white flowers, 1 cm across, are carried along the branches.

HOZAKI-SHIMOTSUKE

SWEET GUMS

Liquidambar formosana   (Hamamelidaceae)   ***(Fū)***

A native of Taiwan and China, though to have been brought from China about 1720. A deciduous tree 8–10 m, seen as a street tree in parks and gardens, and grown particularly for its autumn colour. The sharply pointed 3-lobed leaves, 7–10 cm, are on long stalks and are arranged alternately. The seeds are collected together into balls of about 3 cm that hang down in autumn on stalks 2.5 cm long. In common with many other members of the witch hazel family, the sweet gums produce a scented resin. The autumn colour of the leaves suggests liquid amber as well.

L. styraciflua from Eastern U.S.A. is also grown as a street tree. It is hardier than ***L. formosana*** and colours more brilliantly in the autumn—deep purples, reds and oranges. The leaves are 5 lobed, sometimes 7, with finely toothed edges.

FŪ

MOMIJIBA-FŪ

SWEET POTATO

Ipomoea batatas   (Convolvulaceae)   ***(Satsuma-imo)***

Really a perennial herbaceous plant, but grown as a sprawling annual which is the subject of ongoing hybridization to produce bushier plants with better tubers. The leaves are soft, like pointed hearts on long stalks, the flowers white or pinkish and funnel shaped. The tubers may have white, yellow or orange flesh and red, brown or whitish skins. In Japan they are usually red-skinned.

The northern limit for cultivation is Fukushima. Satsuma is in Kyushu. They were brought from their native Central America by the Spaniards in the hey-day of trade with Spain and Portugal, and have often formed a valuable food crop when rice and wheat have failed. They can be grown on poorer soils and are an important upland vegetable crop, mainly south of 37°. Mostly planted at the beginning of the summer rainy season as a rotational crop.

Bake or steam them as a vegetable. A favourite snack of children. Buy them in autumn and winter from the roast potato man in cities, recognised by his melodious, amplified cries of "***ishi yaki-imo***" (="stone-baked sweet potatoes"). **Shōchū** is the spirit distilled from Japanese sweet potatoes.

SATSUMA-IMO

TARO

Taro, Dasheen, Elephant's Ears *(Sato-imo)*

Colocasia esculenta (Araceae)

The plant grows to 1 m high with characteristic leaves that give it its English name. It will be seen growing commercially and in gardens over most of Japan. The leaf blades are 30–50 cm on long stalks, slightly bent so that the leaf hangs almost vertically. They are light green and thick. Flowers are not often produced. When they are, they are creamy white, hooded and arum-like. It is grown for the edible corms, brown-skinned with white flesh, high in starch, which are an important food crop. They are used in soups and stews.

Cultivars of ***Colocasia*** are also grown as decorative plants for their leaf colour and markings.

SATO-IMO

TEA *(Cha)*

Camellia sinensis (= *Thea sinensis, Camellia Thea*)
(Theaceae)

In its native China and the Himalayas, this is a small tree of 3–4 m. In Japan, it is grown in the Southeast, for example in the Shizuoka region and the Kyoto-Uji area, where the rounded hedges, about 1 m high and 1 m wide, snake around the mountain sides. Really a subtropical evergreen, needing a humid climate and no long exposure to below freezing, it is very similar to the ornamental camellia, and has similar single white flowers. It is the tips of the shoots that are harvested—choice ones by hand, otherwise with mechanical hedgecutters.

O-cha is green tea which, unlike **kōcha**, the usual Western drink, is not fermented or cured, but steamed then immediately dried.

It was originally brought from China by Zen priests for use as a medicine. It gained popularity amongst the upper classes in tea-tasting competitions, and later in the Tea Ceremony. Only during the last 100 years has tea been a drink of the people.

The main types of green tea in daily use are **sen-cha** and **ban-cha**. **Gyokuro** is a finer tea, grown in Uji and Shizuoka where the bushes are shaded from the sun to keep the dark green colour in the leaves. **Sen-cha** is made with water about 90°C, hence the ubiquitous vacuum flask, whereas **gyokuro** needs even cooler water (60°–70°). The new season's shoots (**shincha**) come into the shops in May.

Special varieties are grown for Tea Ceremony tea again with cloth shading over the plants. The young leaves are ground to a powder (**matcha**) before sale. With grace and formality, the tea is mixed in deep bowls with a bamboo whisk, to produce a thick, frothy, bright green tea. A less traditional use is in a pale green, delicately flavoured ice-cream.

Hōji-cha (or **ban-cha**) is made from the last leaves of the season and is "roasted".

CHA

TULIP TREE

(Yuri-no-ki, Hanten-boku)

Liriodendron species (Magnoliaceae)

Liriodendron tulipifera

Identified by the broad deciduous leaves, (6–8 cm), which look as if their tips have been cut off with scissors. What is left is a leaf with a narrow waist and two pointed lobes, which colours bright golden-yellow in the autumn. With imagination, the flowers (in July, greenish yellow, about 6 cm across) can be said to look like tulips, but are really more like magnolias, to which they have a family allegiance. Maybe it is the papery brown fruits in the autumn that are like narrow petalled tulips—again with imagination.

There are trees in Shinjuku-gyoen in Tokyo known to have been planted in the early 1900's that are now 30 m tall.

L. tulipifera is native to Eastern North America and makes a larger tree with leaves which do not have such a narrow waist as in *L. chinense*.

L. tulipfera fastigiatum is a columnar form, ideal for street planting and limited spaces.

YURI-NO-KI

VETCH

Chinese Milk Vetch *(Renge)*

Astragalus sinicus (Leguminosae)

Paddy fields which are squares of beautiful bright purple/pink in spring will be full of this vetch. They are being manured by this biennial plant, with pinnate leaves and a creeping root, that is sown in the autumn. The small pea-shaped flowers are in flat heads about 4 cm across, on stalks 10–30 cm long, looking like little umbrellas. In spring the plant is dug into the land where the nitrogen fixing bacteria in the nodules on the roots form a rich green manure.

WASABI

Japanese Horseradish *(Wasabi)*

Entrema wasabi (=*Wasabia japonica*) (Cruciferae)

A perennial, 30–40 cm high, growing naturally at the edges of streams, it is a typical cruciferous plant with heads of small white 4-petalled flowers followed by long, narrow many-seeded fruits. The leaves are coarse, broad-pointed wedges, and are coarsely toothed. The knobbly, root-like stems, 6–15 cm long and about 2 cm across, are the reason for its cultivation. They have a strong pungent taste that combines well with **shōyu** and **sashimi** (raw fish), with fish **sushi** and with **soba**. The root is a most attractive bright green when it is grated.

RENGE

WASABI

201

WISTERIA

Japanese Wisteria *(Fuji)*

Wisteria floribunda (=*W. multijuga*) (Leguminosae)

Climbs through tall trees in woods all over Honshū. The pale green deciduous leaves have 13–19 leaflets. The scented tassels of pale mauve pea-shaped flowers hang in racemes in May. In gardens they may be as long as 1 m. They open from the top to the bottom. The cultivar *Kyūshaku-fuji* may even have flowers 2 m long. Amongst the many cultivars are *Alba*—white, *Kuchi-beni*—pinkish white, *Macrobotrys*—massive, mauve.

The native wisteria and its cultivars twine clockwise.

Chinese Wisteria

Wisteria sinensis (=*W. chinensis*)

Is a more rampant climber, with shorter leaves, usually 9–13 leaflets, and shorter flower clusters in which the flowers open all at the same time.

It twines anti-clockwise.

Wisteria x formosa is a hybrid of *W. floribunda* and *W. sinensis*.

WITCH HAZEL

Japanese Witch Hazel *(Mansaku)*

Hamamelis japonica (Hamamelidaceae)

A deciduous, open, small tree, 5–6 m high, with spreading branches and broad, pointed, coarsely toothed leaves that colour vivid yellows and reds in the autumn. It grows wild in the mountains all over Japan and is noticeable in winter and early spring for the unusual shaggy, strappy yellow flowers sitting in tight clusters on bare branches.

Chinese Witch Hazel *(Shina-mansaku)*

Hamamelis mollis

Is very similar, but with broader petals and sweetly scented flowers.

There are many cultivars and hybrids of both these species in gardens in Japan, and throughout the cool temperate world.

MANSAKU

ZELKOVA

Zelkova serrata (Ulmaceae) *(Keyaki)*

A native deciduous tree that will grow to 30 m or will make a good bonsai subject. It has a pleasant oval form, and a natural grace, with a firm main trunk and graceful, slender upright, spreading branches suggesting a fan. It is not noted for its flowers or its fruit, which are insignificant, but for its intrinsic graceful symmetry and for its leaves—fresh green in spring, yellows, reds and oranges in the autumn. They are to 5 cm long, sharply toothed (note the specific name), tapering and oval, arranged alternately. The bark on older trees is a characteristic brown, mottled and flaky.

It gives a very fine timber for buildings and utensils, but is very expensive nowadays.

APPENDIX 1

Botanic Gardens

Please note that most of these gardens have an admission fee of ¥100–¥400. Details apply in 1989 but are liable to change.

Botanical Gardens of Tokyo University
Koishikawa Botanic Garden

Tel: (03) 814-0138 Address: 3-7-1, Hakusan, Bunkyo-ku, Tokyo

Open: 9.00 am–4.00 pm

Closed: Mondays and from Dec. 29 to Jan. 3

Access: 15 minute walk from Hakusan Subway Station on the Toei Line, or 15 minute walk from Myogadani Subway Station on the Marunouchi Line. Entrance at S.W. corner of gardens.

A haven for the plantsman. As befits its academic status, it uses scientific names on the plant labels. There are long established collections of plants in a pleasing layout with ponds, woods, greenhouses and beds of collections of herbaceous plants. Interesting at any season, but especially for the glades of plum blossom (*Ume*) in March and the avenue of Maples in Oct./Nov.

National Park for Nature Study

Tel: (03) 441-7176 Address: 5-21-5, Shirogane-dai, Minato-ku, Tokyo

Open: May 1–Aug. 31 9.00 am–4.00 pm
(leave park by 5.00 pm)
Sept. 1–Apr. 30 9.00 am–3.00 pm
(leave park by 4.30 pm)

Closed: Mondays, and on the day following a National Holiday and for about one week around New Year
Access: 10 minute walk from Meguro Station, Yamanote Line, JR

It is amazing to find 20 hectares in the middle of Tokyo that has been undisturbed for centuries. Now it is an organised wilderness and a nature and conservation centre. We give it full marks for its programme of educational classes, exhibitions and guided tours—but for Japanese speakers only. All the plants that you are likely to see on your travels in the countryside are here, and labelled, but strictly for the Japanese. You will need this book.

Shinjuku Gyoen National Garden
Tel: (03) 350-0151 Address: 11 Naitō-chō, Shinjuku-ku, Tokyo

Open: 9.00 am–4.30 pm
Closed: Mondays
Access: 3 minute walk from Shinjuku-gyoen-mae Subway Station on the Marunouchi Line, or 3 minute walk from Sendagaya Station on the JR Sōbu Line, turn right outside the station, then under the railway line and turn left.

Formerly a private mansion, belonging to the State since 1872 and called the Shinjuku Imperial Garden since 1891, it has long been a centre for research and the promotion of horticulture. The present layout was started in 1901, influenced largely by a professor of horticulture from France. This explains the formal bedding of the European Court and the parkland setting for the large collection of foreign trees imported for the first time during the Meiji era. There are also delightful Japanese-style lakes and the largest greenhouse in Asia. There are 59 hectares altogether, with around 1,100 cherry trees of 34 species, and impressive chrysanthemum displays in mid-November. Unfortunately, the trees are only labelled in Japanese.

Kyoto Botanic Garden

Tel: Kyoto (075)-701-0141   Address: Kamohangi-cho, Sakyo-ku, Kyoto

Open: 9.00 am–4.00 pm (leave park by 5.00 pm)

Closed: Dec. 28–Jan. 4

Access: 15 minute walk from Kitaoji Station at the north end of the subway line.

Areas of European formal bedding, mammoth tropical glasshouses, and formalised Japanese ponds give way to more natural lakes and woodlands (24 hectares in all). The representative collections of plants in all these habitats are extensive and the majority of labels use Latin names. Particularly notable are the collections of bamboos, plums, cherries, lotus, medicinal and economic plants. Special exhibitions (e.g. azaleas, orchids) are held throughout the year. Backed by the hills of Kyoto (aflame with Maples in the autumn) and dominated by Mt. Hiei.

Nikkō Botanic Garden

Tel: Nikkō (0288)-54-0206   Address: Hanaishicho 1842 Nikkō. Tochigi Pref

Open: Apr. 15–Nov. 30   9.00 am–4.00 pm (leave gardens by 4.30 pm)

Closed: Mondays

Access: 10 minute walk from Tōshōgu Shrine along the road to Chūzenji. Buses run at frequent intervals from Nikkō railway stations via the main street and pass the Botanic Gardens—details from the Tourist Office in the main street.

An annexe of Koishikawa Botanic Garden and part of the faculty of Science of Tokyo University. A haven for tourists wanting to see something of the natural Japan, and yet within walking distance of a major tourist centre. A chance to see comprehensive collections of cherries, rhododendrons, hydrangeas, magnolias, maples and bush clovers, all named scientifically and in a natural mountain/woodland setting. There are bog and rock gardens, lakes, and a special bonus of a mountain torrent, in its gorge, traversing the gardens. The 10 hectares of land were first developed as an arboretum in 1909, so there are now large trees, making an idyllic retreat for the general public as well as learned botanists.

Matsumae Koen, Hokkaido
Tel: 01394-2-2275 Matsumae Town Office
Address: Matsumae-cho, Matsushiro
Open: Always. The main season is from 25th April to end of May, with a peak about 10th May.
Access: 2 hours 20 minutes by car from Hakodate along Route 228. Follow signs to Matsumae Koen.

A park planted with the intention of bringing together all the *Satozakura* of Japan, including some rare varieties. The collection was started in 1922 and the park was opened to the public in 1960. There is an educational centre (Sakura Mihon-en) with 100 varieties of *Satozakura* and 8,000 other cherry trees in a natural setting. The labels are only in Japanese at the time of writing but we are told that the situation is under review.

Tokyo Yumenoshima Tropical Botanical Garden
Tel: (03) 522-0281
Address: 3-2 Yumenoshima, Koto-ku, Tokyo
Open: 9.30 am–4.00 pm
Closed: Mondays
Access: 7 minute walk from Shin-kiba Subway Station on the Yurakucho Line. Next to Yumenoshima Sports Centre.

Three enormous glass domes house ornamental ferns, a tropical village and a tropical rain forest. There are in all 4,256 plants of 106 species including breadfruit, banana, mango, coffee and palm trees. Species are named in Latin and Japanese.

Other Gardens in Tokyo

Meiji Shrine Inner Garden
Open: 9.00 am–4.00 pm daily
Access: 5 minute walk from Harajuku Station on JR Yamanote Line.
Speciality: Over 100 varieties of Iris in mid June.

East Garden of the Imperial Palace
Open: 9.00 am–3.00 pm (leave gardens by 4.00 pm) When Friday or Monday is a national holiday.
Closed: Friday and Monday. Dec. 25–Jan. 3. When an imperial function is taking place.
Access: At the Ōtemon Gate to the Palace—near the Palace Hotel in the Marunouchi District.
Specialities: Irises, azaleas, pines and cherries. View over the city from the site of the old **honmaru** (donjon) keep of the original palace.

Ueno Park
Open: Always
Access: Adjoining Ueno Station
Specialities: Cherries in April. Lotus in the Shinobazu Pond in August. Site of the National Science Museum, other museums, and the zoo with its Giant Pandas.

Rikugien
Open: 9.00 am–4.00 pm
Closed: Mondays
Access: 8 minute walk from Komagome Station on JR Yamanote Line.

Specialities: Azaleas, carp. 8.8 hectares of typical, and immaculately tended, Japanese landscape (stroll) garden, constructed in the 18th century. With Kōrakuen and Kiyosumi, Rikugien is one of the three most notable landscape gardens in Tokyo.

Kōrakuen

Open: 9.00 am–4.30 pm daily
Closed: Dec. 28–Jan. 4
Access: 7 minute walk from Kōrakuen Subway Station (Marunouchi Line) or 7 minute walk from Iidabashi Station on JR Yamanote Line and Tōzai and Yūrakuchō Subway Lines.
Specialities: Landscape garden, more compact than Rikugien. A 400 year old pine tree and a hill of clipped bamboo. Tranquil oasis amongst the office blocks, the baseball stadium and the amusement park.

Kiyosumi Garden

Open: 9.00 am–4.30 pm daily
Closed: Dec. 29–Jan. 3
Access: 15 minute walk from Monzen-Nakachō Subway Station on the Tōzai Line. In Kōtō-ku.
Specialities: Rocks collected from all parts of Japan, laid out in a landscape/stroll garden in the 19th century.

Hama Rikyū, Detached Palace Garden

Open: 9.00 am–4.30 pm
Closed: Mondays
Access: 15 minute walk from Shimbashi Station
Specialities: On the edge of Tokyo Bay, with a tidal pond. Extensive stroll garden, once an imperial garden and retreat. Clipped pine trees of rare beauty.

Shiba Rikyū, Detached Palace Garden

Open: 9.00 am–4.30 pm
Closed: Mondays
Access: Adjoining Hamamatsuchō Station, JR Yamanote Line. Or get a bird's eye view from the observatory at the top of the World Trade Centre Building—buy a ticket for the elevator on the first floor of the building—access from Hamamatsucho Station.
Specialities: A minor stroll garden that is a miniature oasis in a busy area.

Gardens in Kyoto

These are many. For descriptions, histories and access details we strongly recommend *A Guide to the Gardens of Kyoto* by Marc Treib and Ron Herman, published by Shufunotomo Co., Ltd.

Plant Fairs and Exhibitions in Tokyo

May 31/June 1 June 30/July 1	Asakusa	**Ueki Ichi**, Plant Market, or Potted Tree Fair near Sengen Shrine at north end of Asakusa Kannon Temple
July 6–8	Iriya. Kishibojin Temple	**Asagao Ichi**, Morning Glory Fair near Iriya Subway Station on the Hibiya Line (the next station after Ueno). Thousands of potted Morning Glories on sale.

July 9–10	Asakusa	**Hōzuki Ichi**, Ground Cherry Fair (also known as 'Chinese Lanterns' or *Physalis*). Stalls are set up in the compound of Asakusa Kannon Temple
Mid-October–Mid-November	Shinjuku-gyoen Yasukuni Shrine Meiji Shrine Asakusa Kannon Temple and Hibiya Park	Chrysanthemum Shows

APPENDIX II

The Seven Herbs of Spring *(Haru no nanakusa)*

These 7 plants, showing green through the cold earth, were regarded as harbingers of spring in *the Manyōshū*, the oldest anthology of Japanese poetry composed in the 400 years up to 759 AD. Traditionally the leaves were gathered on the seventh day of the New Year, chopped and added to rice gruel. It is still the custom to eat this on the morning of January 7, not so much for its medicinal effect, but as an antidote to the rich food of the New Year. Bunches of the following herbs are on sale at supermarkets:

Seri	Water Parsley	*Oenanthe stolonifera*
Nazuna	Shepherd's Purse	*Capsella bursa-pastoris*
Gogyō	Cudweed	*Gnaphalium multiceps*
Hakobe	Chickweed	*Stellaria media*
Hotokenoza	Henbit, Deadnettle	*Lamium amplexicaule*
Suzuna (Kabu)	White Turnip	*Brassica campestrus subsp. Rapa*
Suzushiro (Daikon)	Japanese White Radish	*Raphanus sativus var. hortensis*

The Seven Flowers of Autumn *(Aki no nanakusa)*

These 7 flowers were chosen in *the Manyōshū* to correspond to the 7 herbs of spring. Of all the autumn flowers, they were thought to express most clearly the melancholy of autumn.

Hagi	Bush Clover	*Lespedeza spp.*
Obana (Susuki)	Japanese Pampas Grass	*Miscanthus sinensis*
Kuzu	Kuzu Vine	*Pueraria Thunbergiana*

Nadeshiko Fringed Pink *Dianthus superbus longicalycinus*
Ominaeshi *Patrinia scabiosaefolia*
Fujibakama *Eupatorium stoechadosmum*
Asagao Morning Glory *Pharbitis nil*
 (though this was most probably *Kikyō*, Balloon Flower, *Platycodon grandiflora*)

APPENDIX III

Classification of Cherry Blossoms by courtesy of
Prof. Ryuzo Sakiyama

Wild Cherries
The following classification, proposed by Hayashi and Tominari in *Ornamental Trees and Shrubs of Japan* (1971), is generally recognized in Japan at present.

A. Subgen cerasus

  I. *Yama-zakura Group*
   1. *Yama-zakura* — *P. jamasakura* Sieb, ex Koidz
   2. *Ōyama-zakura* — *P. sargentii* Rehder
   3. *Mame-zakura* — *P. incisa* Thunb
   4. *Kasumi-zakura* — *P. verecunda* (Koidz.) Koehne
   5. *Ōshima-zakura* — *P. lannesiana* (Carr.) Wilson var. *speciosa* (Koidz.)
   6. *Miyama-zakura* — *P. maximowiczii* Rupr.
   7. *Mine-zakura* — *P. nipponica* Matsum.

  II. *Higan-zakura Group*
   8. *Edo-higan* — *P. pendula* Maxim. *f. ascendens* Chwi

  III. *Choji-zakura Group*
   9. *Choji-zakura* — *P. apetala* (Sieb. et. Zucc) Franch. et. Savat

B. Subgen padus
  I. *Inu-zakura* — *P. buergeriana* Miqvel
  II. *Uwamizu-zakura* — *P. grayana* Maxim.

Cultivated Cherries
Somei-Yoshino, Sato-zakuras (Shidare-zakura, Yae-zakuras, etc.)

The *Yama-zakura* group of Wild Cherries has had the greatest effect on the formation of the cultivated cherries. Most of the important cultivars were derived from *Ōshima-zakura*, with the influence of *Yama-zakura, Ōyama-zakura, Kasumi-zakura, Edo-higan* or *Choji-zakura*. Cultivated cherries found near old Tokyo (Edo) were much related to *Ōshima-zakura*. (Ōshima is an island in the Pacific Ocean 100 km from Tokyo.) In the Kansai district many cultivars originated from *Yama-zakura* or under its influence.

Before the Meiji era, when the word *Sakura* was used by the people it referred to *Yama-zakura*. After the Meiji era *Sakura* has usually meant *Somei-Yoshino*.

BIBLIOGRAPHY

Aoki & Masui. *Ume 12 ka Getsu*. Shufunotomo, 1982

Austin, Ueda & Levy. *Bamboo*. Weatherhill, 1970

du Cane, Florence. *The Flowers and Gardens of Japan*. A & C Black, 1908

Darden, Jim. *Great American Azaleas*. Greenhouse Press, 1985

Flower Association of Japan. *Manual of Japanese Flowering Cherries*, 1983

Hillier. *Color Dictionary of Trees and Shrubs*. David & Charles, 1981

Joya, Mock. *Things Japanese*. Japan Times Ltd, 1985

Kitanokoji, Isamitsu. *Cherry Blossoms; Japan in the Springtime*. Kodansha International, 1973

Makino, Tomitaro. *New Illustrated Flora of Japan*. Hokuyukan, 1963

Nakajima, Young. *The Art of the Chrysathemum*. Harper & Row, 1965

Numata, N. *The Flora and Vegetation of Japan*. Kodansha International, 1974

Nuttonson, M.Y. *Ecological Geography and Crop Practices of Japan*. American Institute of Crop Ecology, 1951

Ohwi, J. *Flora of Japan*. Shibundo, 1956

Okita & Hollenberg. *The Miniature Palms of Japan*. Weatherhill, 1981

Takayanagi, Y. *Masterpieces of Bonsai*. Shufunotomo, 1986

Tominari, Tadao. *Jumoku*. Yama-kei, 1979

Treib & Herman. *Guide to the Gardens of Kyoto*. Shufunotomo, 1981

Wright, M. *Complete Handbook of Garden Plants*. Michael Joseph, 1984

INDEX

Acer 128, 130
Adonis 8
Ai 102
Ajisai 100
Aka-jiso 178
Aka-matsu 160
Akame-mochi 156
Akebono-sugi 76
Albizia 180
Amacha 100
Ama-gaki 154
American plane 164
Amerika suzukake-no-ki 164
Amorphophallus 112
An 32
Andromeda 84, 158
Anzu 14
Aoki 16
Ao-jiso 178
Apples 10
Apricots 12, 14
Aralia 88
Arctium 38
Ardisia 16
Arundinaria 24, 26
Asagao 134
Asarum 146
Asebi 158
Astragalus 200
Aucuba 16
Ayame 104
Azaleas 18
Azuki 32
Azukia 32
Azuki beans 32

Balloon flower 36
Bamboos 24
Ban-cha 196

Bandai-sugi 48
Barberries 30
Beafsteak plant 178
Beans 32
Beauty berry 36
Bellflower 36
Benihana 174
Berberis 30
Biwa 120
Boke 168
Bog rhubarb 42
Bonsai 60, 128, 130, 160
Botan 140
Buckwheat 38
Buntan 66
Burdock 38
Bush clover 40
Bushido 54
Butterbur 42
Buttonball 164
Buttonwood 164

Callicarpa 36
Calystegia 134
Camellias 44
Camellia 44, 196
Campanula 36
Camphor tree 46
Carthamus 174
Cedar, Japanese 48
Cedar, white 72
Cercidiphyllum 108, 110
Cercis 108
Cha 196
Chamaecyparis 72
Chaenomeles 168
Chawan mushi 92
Cherry blossoms 50
China berry 102

222

Chinese milk vetch   200
Chinese pieris   158
Chinese redbud   108
Chinese tulip tree   198
Chinese windmill palm   144
Chinese witch hazel   204
   Chloranthus   176
Chrysanthemums   58, 60, 62
   Chrysanthemum   58
Chū-giku   62
Chusan palm   144
   Cinnamomun   46
   Citrus   64, 68
Citrus fruits   64
   Cleyera   86
   Cornus   82
   Colocasia   194
Coral berry   16
Crab apples   10
   Cryptomeria   48
Cucumber tree   126
Cycads   70
   Cycas   70
   Cydonia   168
Cypresses   72

Daikon   74
Daizu   32
   Daphne   74
Dasheen   194
Dwarf palms   146
Dawn redwood   76
Day lily   78
Dengaku   174
   Dianthus   80
   Diospyros   154
Dōdan-tsutsuji   84
Dog's tooth violet   80
Dogwoods   82
Double cherries   52
Downy Japanese maple   128

Ego-no-ki   182
Elephants' ears   194

Enju   142
   Enkianthus   84
Equinox flower   184
   Equisetum   132
   Eriobotrya   120
   Erythronium   80
   Eulalia   94
   Eurya   86

   Fagopyrum   38
False castor oil plant   88
   Fatsia   88
   Fortunella   66
Fossil tree   76
Fringed pink   80
Fuji   202
Fuki   42
Fukujusō   8
Fū   190

Gaku-ajisai   100
   Gardenia   88
Geta   126, 148
Ginger   90
   Ginkgo   92
Gin-mokusei   138
Ginnan   92
   Glycine   32
Gobō   38
Gohan   172
Goyō-matsu   160
Grasses   94
Green shiso   178
Grey-budded snake bark maple   130
Gyokuro   196

Hagi   40
Haku-mokuren   126
Hakuunboku   182
   Hamamelis   10
Hana-kaidō 10
Hana-shōbu   104
Hanten-boku   198
Hanami   54

223

Nikkō-kisuge 78
O-cha 196
Oden 112
Ogi 94
Oni-yuri 116
Omoto 146
Oranges 64, 68
Oroshima-chiku 26
    Oryza 172
    Osmanthus 138
Otome-yuri 118
Ō-yamazakura 50

Paeonies 140
    Paeonia 140
Pagoda tree 142
Palms 144, 146
Pampas grass 94
    Paulownia 148
Peach 150
Pears 152
    Perilla 12, 178
Persian Acacia 180
Persian lilac 102
    Petasites 42
    Pharbitis 134
    Phaseolus 32
    Photinia 156
    Phyllostachys 24
    Pieris 158
Pines 160
    Pinus 160, 162
Pinks 80
Pink siris tree 180
Planes 164
    Platanus 164
    Platycodon 36
Plums 12
    Podocarpus 166
    Polygonum 102
    Poncirus 68
    Prunus 12, 50, 52, 150
    Pyrus 152
Pomelo 66

Quince 168

    Raphanus 74
Redbud 108
Reeds 170
Renge 200
Renge-tsutsuji 20
Renkon 122
    Retinospora 72
    Rhapis 146
    Rhizobium 32
    Rhododendron 18
    Rhus 112
Rice 172
Rindai 60
Ringo 10
    Rohdea 146
Rushes 170
Ryō 16
Safflower 174
Sakaki 86
    Sakakia 86
Sake 172
Sakura 50
Sakura mochi 56
Sansho 174
Sarasa-dōdan 84
Sasa 24, 26
    Sasa 24
Sasa-yuri 118
Sato-imo 194
Sato-zakura 52, 54
Satsuki 18
Satsuma-imo 192
Satsuma orange 64
Sawara 72
Sawara cypress 72
Sazanka 44
    Sciadopitys 162
Sekihan 32
Seiyō-hanazuo 108
Seiyō-nashi 152
Sen-cha 196
Sendan 102

Senryō 176
Shabu shabu 34
Shaga 106
Shakunage 18
Shakuyaku 140
Shamisen 54
Shibu-gaki 154
Shidare-momo 150
Shidare-zakura 52
Shide-kobushi 124
Shimotsuke 186
Shin-cha 196
Shirataki 112
Shiso 12, 34, 178
Shōga 90
Shōga-yu 90
Shōchu 14
Shōyu 28, 34, 200
Shun-giku 58
Shuro 144
Shuro-chiku 146
Silk tree 180
Silver grass 94
Snowbell tree 182
Soba 38, 200
Somei-yoshino 50, 52
Sophora 142
Sotetsu 70
Soya bean 32, 34
Spider lily 184
Spiraea 186, 188
Spotted laurels 16
Storax 182
Strawberry tree 82
Styrax 182
Sugi 48
Sukashi-yuri 118
Sukiyaki 34, 58, 112
Sumomo 14
Sushi 74, 154, 178, 200
Susuki 94
Suzukake-no-ki 164
Sweet daphne 74
Sweet gums 190

Sweet olive 138
Sweet potato 192

Takanoha-susuki 94
Take 24
Take-no-ko 28
Taro 194
Tatami 170
Tatami reed 170
Tea 44, 196
Tempura 80, 178
Teppō-yuri 116
Tiger lily 116
Tōfu 34, 178
Tōjuro 144
      Trachycarpus 144
Tsubaki 44
Tsukushi 132
Tsutsuji 18
Tulip tree 198

Ukon 52
Umbrella pine 162
Ume 12
Ume-boshi 12, 178
Ume-shu 14
Unshū-mikan 65
Urushi 112

Vetch 200
    Vigna 32

Wasabi 200
    Wasabia 200
Weeping cherry 52
Western pears 152
    Wisteria 202
Witch hazel 204

    Xanthoxylum 174

Yabu-kanzō 78
Ya-dake 26
Yae-zakura 52

227

Yamabōshi 82
Yamabuki 110
Yama-tsutsuji 18, 22
Yama-yuri 116
Yama-zakura 50
Yatsude 88
Yuki-yanagi 186
Yuri-no-ki 198
Yuzu 68

Zabon 66
Zanthoxylum 174
Zelkova 206
Zingiber 90
Zoysia 96